T0185628

Advanced Courses in Mathematics - CRM Barcelona

Managing Editor

David Romero i Sànchez, Centre de Rercerca Matemàtica, Barcelona, Spain

Since 1995 the Centre de Recerca Matemàtica (CRM) has organised a number of Advanced Courses at the post-doctoral or advanced graduate level on forefront research topics in Barcelona. The books in this series contain revised and expanded versions of the material presented by the authors in their lectures.

Colin Christopher • Chengzhi Li
Joan Torregrosa

Limit Cycles of Differential Equations

Second Edition

 Birkhäuser

Colin Christopher
School of Engineering, Computing and Mathematics
University of Plymouth
Plymouth, Devon, UK

Chengzhi Li
School of Mathematical Sciences
Peking University
Beijing, China

Joan Torregrosa
Departament de Matemàtiques
Universitat Autònoma de Barcelona
Bellaterra, Barcelona, Spain

ISSN 2297-0304   ISSN 2297-0312   (electronic)
Advanced Courses in Mathematics - CRM Barcelona
ISBN 978-3-030-59655-2   ISBN 978-3-030-59656-9   (eBook)
https://doi.org/10.1007/978-3-030-59656-9

This book is published under the imprint Birkhäuser, www.birkhauser-science.com by the registered company Springer Nature Switzerland AG
The registered company address is: Gewerbestrasse 11, 6330 Cham, Switzerland

Paper in this product is recyclable.

Foreword

This book contains two sets of revised and augmented notes prepared for the Advanced Course on Limit Cycles and Differential Equations given at the Centre de Recerca Matemàtica in June 2006, as part of its year-long research programme on Hilbert's 16th problem. The common goal of the two sets of notes is to help young mathematicians enter a very active area of research lying on the borderline between dynamical systems, analysis and applications.

The first part of the book, by Colin Christopher, considers some of the topics which surround the Poincaré center-focus problem for polynomial systems, a subject closely tied with the integrability of polynomial systems. The second part, by Chengzhi Li and Joan Torregrosa, is devoted to the introduction of some basic concepts and methods in the study of Abelian integrals and applications to the weak Hilbert's 16th problem.

Besides our indebtedness to the Centre de Recerca Matemàtica, thanks are due to Jaume Llibre and Armengol Gasull, the course co-ordinators, for giving us this challenging but rewarding opportunity and for providing such a pleasant environment during the programme.

Contents

I   Around the Center-Focus Problem
Colin Christopher   1

Preface   3

1   Centers and Limit Cycles   5
   1.1   Outline of the Center-Focus Problem   5
   1.2   Calculating the Conditions for a Center   10
   1.3   Bifurcation of Limit Cycles from Centers   13

2   Darboux Integrability   21
   2.1   Invariant Algebraic Curves   21
   2.2   The Darboux Method   22
   2.3   Multiple Curves and Exponential Factors   25

3   Liouvillian Integrability   31
   3.1   Differential Fields and Liouvillian Extensions   31
   3.2   Proof of Singer's Theorem   32
   3.3   Riccati equations   36

4   Symmetry   39
   4.1   Algebraic Symmetries   39
   4.2   Centers for analytic Liénard equations   41
   4.3   Centers for polynomial Liénard equations   43

5   Cherkas' Systems   49

6   Monodromy   59
   6.1   Some Basic Examples   59
   6.2   The Model Problem   61
   6.3   Applying Monodromy to the Model Problem   62

7   The Tangential Center-Focus Problem   **69**
     7.1   Preliminaries   71
     7.2   Generic Hamiltonians   72
     7.3   Relative exactness   73

8   Monodromy of Hyperelliptic Abelian Integrals   **79**
     8.1   Some Group Theory   80
     8.2   Monodromy groups of polynomials   81
     8.3   Proof of the theorem   83
     8.4   The symmetry of the differential   85

9   Holonomy and the Lotka–Volterra System   **89**
     9.1   The monodromy group of a separatrix   90
     9.2   Integrable points in Lokta–Volterra systems   92
     9.3   Holonomy on more general curves   96

10 Other Approaches   **99**
     10.1 Finding components of the center variety   99
     10.2 Extending Centers 101
     10.3 An Experimental Approach 103

Bibliography   **107**

**II   Abelian Integrals and Applications to the Weak
       Hilbert's 16th Problem**
       Chengzhi Li & Joan Torregrosa   **117**

Preface of the first edition   **119**

Preface of the second edition   **121**

1   Hilbert's 16th Problem and Its Weak Form   **123**
     1.1   Hilbert's 16th Problem 123
           1.1.1   The finiteness problem 124
           1.1.2   Configuration of limit cycles 124
           1.1.3   Some results on quadratic systems 125
           1.1.4   Some results on cubic and higher degree systems 127
           1.1.5   Some results on Liénard equations 129
     1.2   Weak Hilbert's 16th Problem 130
           1.2.1   The study of $\tilde{Z}(2)$ 131
           1.2.2   Perturbations of elliptic and hyperelliptic Hamiltonians . . 133

2 Abelian Integrals and Limit Cycles **143**
   2.1 Poincaré–Pontryagin Theorem 143
   2.2 Higher Order Approximations 148
   2.3 The Integrable and Non-Hamiltonian Case 155
   2.4 The Study of the Period Function 161

3 Estimate of the Number of Zeros of Abelian Integrals **167**
   3.1 The Method Based on the Picard–Fuchs Equation 167
   3.2 A Direct Method 170
   3.3 The Method Based on the Argument Principle 176
   3.4 The Averaging Method 182
   3.5 The Averaging Method in Piecewise Systems 187
   3.6 Other Methods and Related Works 190

4 A Unified Proof of the Weak Hilbert's 16th Problem for n=2 **193**
   4.1 Preliminaries and the Centroid Curve 193
   4.2 Basic Lemmas and the Geometric Proof of the Result 196
   4.3 The Picard–Fuchs Equation and the Riccati Equation 199
   4.4 Outline of the Proofs of the Basic Lemmas 206
   4.5 Proof of Theorem 4.6 207

Bibliography **211**

Part I

Around the Center-Focus Problem

Colin Christopher

Preface

My aim in these notes is to consider some of the topics which surround the Poincaré center-focus problem for polynomial systems. That is, given a polynomial system

$$\dot{x} = P(x, y), \qquad \dot{y} = Q(x, y),$$

with P and Q polynomials, which has a critical point whose linearization gives a center, under what conditions can we conclude that the point is a center for the nonlinear system?

Clearly, the subject is closely tied with understanding what mechanisms underlie the local integrability of polynomial systems, since the existence of a center implies the existence of a local analytic first integral.

Because the defining systems are algebraic, we would expect these mechanisms to be algebraic too in some sense. This indeed seems to be the case, but the situation is far from being well understood except for a growing number of explicit examples.

The choice of topics covered in these notes is very much a personal one, being in the main problems that I have been involved in myself or found interesting. Unfortunately, this has meant that there is much that is missing from this presentation which I felt less competent to comment on. In particular, very little is said on the many detailed analyses of particular systems, nor on the more far-reaching work on holomorphic foliations.

The first part of the notes considers the two main mechanisms known to produce centers in polynomial systems, namely Darboux integrability and algebraic symmetries. The second part considers several topics loosely associated with the idea of monodromy. Though diverse, they share a common theme of teasing out concrete global information from trying to extend the known local behavior, surely one of the most beguiling aspects of the center-focus problem.

For this second edition I have added some clarifications where necessary to the main sections and updated the notes substantially to reflect some new modern developments. I have also added some questions for future research.

During the revision of these notes, Noel Lloyd (1946-2019) sadly passed away after a long battle with cancer. Much of the initial work using computer algebra systems to compute small-amplitude bifurcations came through his own researches and that of his group. I would like to dedicate these notes to his memory.

Chapter 1

Centers and Limit Cycles

In this chapter I want to give a general background to the center-focus problem, and then to show why the problem is interesting: both in what it tells us about the distinctive algebraic features of polynomial vector fields, and also in the simple concrete estimates it gives of the number of limit cycles which can exist in these vector fields.

1.1 Outline of the Center-Focus Problem

Let X be a polynomial vector field

$$X = P(x,y)\frac{\partial}{\partial x} + Q(x,y)\frac{\partial}{\partial y}, \tag{1.1}$$

where P and Q are real polynomials of degree at most d. We will identify this vector field with the pair of first-order differential equations,

$$\dot{x} = P(x,y), \qquad \dot{y} = Q(x,y). \tag{1.2}$$

We are interested in the situation where this vector field has a critical point which we can choose, without loss of generality, to be at the origin.

The associated linearized system at the origin is given by calculating the Jacobian matrix $J_{(0,0)}$ where

$$J_{(x,y)} = \begin{pmatrix} \frac{\partial P}{\partial x} & \frac{\partial P}{\partial y} \\ \frac{\partial Q}{\partial x} & \frac{\partial Q}{\partial y} \end{pmatrix}.$$

Then

$$\begin{pmatrix} \dot{x} \\ \dot{y} \end{pmatrix} = J_{(0,0)} \begin{pmatrix} x \\ y \end{pmatrix} + O(2),$$

where $O(2)$ represents terms of degree 2 or higher in x and y.

© The Author(s), under exclusive license to Springer Nature Switzerland AG 2024
C. Christopher et al., *Limit Cycles of Differential Equations*, Advanced Courses
in Mathematics - CRM Barcelona, https://doi.org/10.1007/978-3-030-59656-9_1

If the determinant of $J_{(0,0)}$ is non-zero (we say the critical point is *non-degenerate* in this case), then the Hartman–Grossman theorem tells us that in a sufficiently small neighborhood of the origin, the system is topologically equivalent to its linear part (i.e. we can ignore the terms of higher order) as long as the eigenvalues of $J_{(0,0)}$ are not pure imaginary. That is, as long as the linear parts do not give a center, the Jacobian characterises the topological type of the critical point and the nonlinear system is topologically equivalent to its linearization. This result also holds when P and Q are just continuously differentiable.

If the eigenvalues of $J_{(0,0)}$ are pure imaginary, then the linear part of the system can be brought into the form

$$\dot{x} = -ky, \qquad \dot{y} = kx,$$

after a change of coordinates, where k is some non-zero constant. The trajectories of this system are just the circles $x^2 + y^2 = c$. That is, there is a family of periodic orbits around the origin. When we add the nonlinear terms, however, these closed orbits can be broken and the system will then have a stable or unstable (weak) focus.

The case above is the generic situation. However, the nonlinear terms could also be such that the critical point of the original system is still surrounded by a family of periodic orbits and the center is preserved. This is exactly the case we want to explore in these notes.

The *center-focus problem* asks for the criteria which determine whether a critical point whose linear parts give a center, really is a center.

Note that if we are only interested in whether a critical point is a focus or a center we can perform a scaling of the time, $dt' = k\,dt$, to bring the linear part given above to the form

$$\dot{x} = -y, \qquad \dot{y} = x\,.$$

More generally, the nonlinear timescaling, $dt' = k(x,y)dt$, with $k(0,0) \neq 0$, will give the same topological behaviour for the trajectories, but changes the system (1.2) to the system

$$\dot{x} = P(x,y)/k(x,y), \qquad \dot{y} = Q(x,y)/k(x,y). \tag{1.3}$$

We will often make use of such transformations. Note, for the case of centers, we do not usually specify that $k(0,0) > 0$ so that the topological equivalence will include possible time reversal. If we were interested in the stability or otherwise of the pertubed foci or limit cycles, then we clearly would need this as an additional condition.

If the critical point is either a center or focus, we shall use the more general term *monodromic* to cover both cases. A focus whose linearization gives a center is called a *weak focus*. The following proposition is straightforward from the Hartman–Grossman theorem.

Proposition 1.1. *Suppose that the polynomial system* (1.2) *has a non-degenerate critical point at the origin. If the critical point is monodromic, then we can bring the vector field to the form*

$$\dot{x} = -y + \lambda x + p(x, y), \qquad \dot{y} = x + \lambda y + q(x, y), \tag{1.4}$$

by a linear transformation and a time scaling, where p *and* q *are polynomials without constant or linear terms. The case when* $\lambda = 0$ *corresponds to a weak focus or a center.*

Unless otherwise stated, we take our polynomial system in the form (1.4) with p and q polynomials of degree at most n.

Example 1.2. The linear parts of the system

$$\dot{x} = -y + x^3, \qquad \dot{y} = x + y^3, \tag{1.5}$$

about the origin give a center, but for the nonlinear system we have

$$\frac{d}{dt}(x^2 + y^2) = 2(x^4 + y^4), \tag{1.6}$$

and so trajectories travel away from the origin, and the system has therefore an unstable focus there.

For differentiable systems, the behavior at the origin can be hard to determine as the following well-known example shows.

Example 1.3. The C^∞ system

$$\dot{x} = -y + xf(x, y), \qquad \dot{y} = x + yf(x, y), \tag{1.7}$$

with

$$f(x, y) = \sin\left(\frac{1}{x^2 + y^2}\right) e^{-1/(x^2 + y^2)},$$

has an infinite number of limit cycles, $x^2 + y^2 = 1/n\pi$, for $n \in \mathbb{Z}_+$ accumulating at the origin.

However for polynomial (or analytic) systems, this situation does not occur. A critical point whose linear parts give a center is either asymptotically stable, asymptotically unstable or it is a center. This can be most easily seen by computing the return map at the origin.

That is, we choose a one-sided analytic transversal, Σ, at the origin (see Figure 1.1) with a local analytic parameter c, and represent the return map by an expansion

$$c \mapsto h(c) = \sum_{i=1}^{\infty} \alpha_i(\theta) c^i, \tag{1.8}$$

By expressing (1.4) in polar coordinates,

$$\dot{r} = \lambda r + O(r^2), \qquad \dot{\theta} = 1 + O(r), \tag{1.9}$$

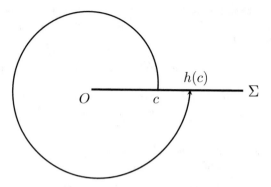

Figure 1.1: Defining the return map.

and invoking standard theorems on analytic dependence on parameters for solutions of the system (1.9), we see that the map (1.8) is analytic in c and also in the parameters of the system, so the expansion (1.8) is valid.

The stability of the origin is clearly given by the sign of the first non-zero α_i, and if all the α_i are zero, then the origin is a center.

Furthermore, in this latter case, we can construct an analytic first integral for the system of the form $x^2 + y^2 + \cdots$ (see below).

In general, it is easy to see that

$$\alpha_1 = e^{2\pi\lambda} - 1 = 2\pi\lambda + O(\lambda^2),$$

and hence the first order term of the return map determines the stability of the linearized system. That is, if λ is non-zero, we have a strong focus.

However, we can say more. The terms α_{2k} are just analytic functions (with zero constant term) of the previous α_i, so the first non-zero coefficient of the return map, if it exists, must be one of the α_{2i+1}.

If α_{2k+1} is the first non-zero one of these, then at most k limit cycles can bifurcate from the origin. We call this a *weak focus of order k*. Provided we have sufficient control of the coefficients α_i, we can also obtain k limit cycles in a simultaneous bifurcation from the critical point. These are sometimes referred to as *small amplitude* limit cycles to distinguish them from limit cycles arising from non-local bifurcations.

We call the functions α_{2i+1} the *Lyapunov quantities* of the critical point, and denote them by $L(i)$. If all the $L(i)$ vanish, then the critical point clearly is a center. When $\lambda = 0$, the $L(i)$ turn out to be polynomials in the coefficients of the monomials of the system (1.4). By the Hilbert basis theorem, the vanishing of all the $L(i)$ must be equivalent to the vanishing of the first N of them, for some integer N.

Thus, if we consider the general case of system (1.4) where the coefficients of the polynomials p and q are the parameters of the system, the set of points where we have a center must be an algebraic set, which we call the *center variety*.

It would appear that one part of the center-focus problem is therefore quite easy, as the calculation of the $L(i)$ is computationally straightforward and has been implemented by many authors. Although these expressions can be large and unwieldy as the number of parameters increase, they are in theory manageable by a computer algebra system.

However, this is deceptive in two ways. First, because the actual calculation of the common zeros of the first N Lyapunov quantities is computationally intensive (and in general intractable for even quite simple systems), and second because the Hilbert Basis Theorem gives us no explicit value for N. That is, we have no idea in practice when to stop calculating values of $L(i)$.

In order to remedy the second problem, we need to know what mechanisms in polynomial systems force the origin to be a center, then we can show that a particular set of parameter values do indeed give a center. Here lies the main interest in the center-focus problem. This is because these mechanisms should reflect something of the algebraic nature of the systems in which they arise. And indeed, this seems to be the case, at least for the families of systems whose centers have been classified to date. In contrast, a generic finite dimensional family of analytic systems of the form (1.4) will have only trivial centers, because the existence of a center has an arbitrarily high codimension generically.

It is conjectured that there are only two main mechanisms which underlie the existence of a center. One is the existence of enough invariant algebraic solutions that an integrating factor can be constructed from them; we consider this case in the next chapter. That is, we seek a first integral or an integrating factor of the form

$$e^{g/h} \prod_i f_i^{l_i},$$

where f_i, f and g are polynomials, and $f_i = 0$ and $h = 0$ define invariant algebraic curves in system (1.4). We call such a function a *Darboux* function, and the center a *Darboux center*. The function $e^{g/h}$ arises when two or more invariant algebraic curves coalesce at a particular parameter value. The resulting *exponential factor* (also known as a *Darboux factor* or *degenerate invariant algebraic curve*) will be considered in more detail in Section 2.3.

The second mechanism is the existence of an algebraic symmetry, that is a map $(x, y) \mapsto (X(x, y), Y(x, y))$, where X and Y are algebraic functions of x and y, which keeps the system fixed but "reverses" time. Any critical points which lie on the set of fixed points of this transformation will be forced to be centers. We consider this case in more detail in Chapter 4. Both these mechanisms clearly have important global consequences for the systems which exhibit them.

We shall see below that the calculation of the Lyapunov quantities can be made purely algebraic, and their vanishing corresponds to the algebraic fact of the existence of a formal power series ϕ such that $X(\phi) = 0$. Seen in this light, the center-focus problem becomes an algebraic question of showing that the existence of a formal local first integral implies the global existence of algebraic solutions or symmetries. It is this fascinating, and unobvious, connection between the local and global properties of polynomial systems that underlies part of the fascination of the center-focus problem.

The other spur to understand what underlies centers in more detail is that they seem a to be a natural "organizing center" for the dynamics of polynomial systems. Due to their algebraic structure, they are also much easier to analyze by perturbation methods. Indeed, many of the strongest conclusions about Hilbert's 16th problem on the number of limit cycles of (1.2) have come exactly from analyzing bifurcations from centers. We give examples of this at the end of this chapter.

With this background in hand, the main conjecture for the center focus problem was first formally stated by Żołądek. I have replaced the original *rationally reversible* by the more general algebraic symmetries as it seems we really do need these in more complex examples.

Conjecture 1.4 (Żołądek). Suppose (1.2) has a center; then the center is either Darboux, or arises from an algebraic symmetry.

In the case of integrable saddles there is also another type of integrability condition when the system can be reduced to a Riccati equation, and hence the solutions expressible in terms of second order linear equations. We give an example in Section 3.3.

1.2   Calculating the Conditions for a Center

In practice, the computation of the Lyapunov quantities from the return map $h(c)$ is not the most efficient way to proceed. Instead, we use a method which turns out to be equivalent. It is clear that to find a center, we only need to calculate the Lyapunov quantities $L(k)$ modulo the previous $L(i)$, $i < k$. In particular, $L(0)$ is a multiple of λ and so we can assume that $\lambda = 0$ when we calculate the $L(k)$ for $k > 0$.

We seek a function $V = x^2 + y^2 + \cdots$ such that for our vector field

$$X = (-y + p(x,y))\frac{\partial}{\partial x} + (x + q(x,y))\frac{\partial}{\partial y},$$

we have

$$X(V) = \eta_4(x^2 + y^2)^2 + \eta_6(x^2 + y^2)^3 + \cdots, \tag{1.10}$$

for some polynomials η_{2k}. The calculation is purely formal, and the choice of V can be made uniquely if, for example, we specify that $V(x,0) - x^2$ is an odd function. It turns out that the polynomials η_{2k+2} for $k > 0$ are equivalent to $L(k)/\pi$ modulo the previous $L(i)$ with $i < k$.

This can be seen by considering polar coordinates, as we did before for the return map, and taking

$$\rho = \sqrt{V(r\cos\theta, r\sin\theta)} = r + O(r^2),$$

to give

$$\dot{\rho} = \eta_{2k+2}\rho^{2k+1}/2 + O(\rho^{2k+2}),$$

where η_{2k} is the first non-vanishing of the η_{2i}. Calculation of the return map in this new coordinate system gives the result.

Though this is a purely algebraic way to calculate the Lyapunov quantities, it turns out that, if the origin is a center, then the expression for V converges to an analytic function. Thus the existence of a formal first integral V and an analytic one are equivalent in this case. This justifies our assertion that the center-focus problem is a purely algebraic phenomenon.

If the linear parts of the system are not quite in the form of (1.4), then rather than transform the system to (1.4), we can replace the terms $x^2 + y^2$ in expansion of V by the equivalent positive definite quadratic form which is annihilated by the linear parts of the vector field.

We note that if we have a center at the origin with first integral V as above, we can always choose coordinates X and Y such that $V = X^2 + Y^2$. The system is thus orbitally equivalent to the linear center:

$$\dot{X} = -Y, \qquad \dot{Y} = X. \tag{1.11}$$

That is, the system can be brought to this form after multiplying by some analytic function $h(X,Y)$ with $h(0,0) \neq 0$.

Thus, from the analytic point of view all centers are equivalent. It is only as we restrict our attention to algebraic phenomena that we see the richness of the various center types.

We mention, to close this section, a closely related phenomena for which many of the techniques in these notes can be applied.

The system (1.11) has a constant period for each periodic orbit. In general, we say that a center is *isochronous* if it has a constant period for all the periodic orbits in a neigbourhood of the center.

Clearly this definition is independent of changes of coordinates, which can be chosen arbitrarily, but will not allow non-linear scalings of time as we have used above. This gives the mecahnisms surrounding of isochronous centres a quite different flavour, in some ways, from center conditions.

If we do not allow a scaling of time, we can bring any center to the form (1.11) with a non-constant factor:

$$\dot{X} = -Yg(X,Y), \qquad \dot{Y} = Xg(X,Y), \qquad g(0,0) \neq 0. \tag{1.12}$$

The period of the orbits, $r = c$, are given by the *period function*

$$T(c) = \int_0^{2\pi} \frac{d\theta}{g(c\cos(\theta), c\sin(\theta))}. \tag{1.13}$$

This too can be developed in a series in c:

$$T(c) = a_0 + \sum_{i=1}^{\infty} a_i c^i, \qquad a_0 = 2\pi/g(0,0).$$

convergent for c sufficiently small. Expanding the integrand of (1.13) , we see that the terms of the form a_{2i+1} vanish. The quantities a_{2i} are called the *period constants* of the center. If the period constants $a_{2i} = 0$, $i = 1, ..., k-1$ with $a_{2k} \neq 0$, then a perturbation of the center within the family of centers can generate at most k turning points of the function $T(c)$ for c in an interval $(0, \epsilon)$ with ϵ sufficiently close to zero. These turning points are called critical periods.

The theory of the bifurcation of critical periods follows very similar lines to the considerations we will make here and has been investigated by several authors starting with [36] as well as the question of the global behaviour of the period function [92].

If the system, or its perturbations, do not have a center, we still can make some sense of the concept of critical period, as long as the critcal point remains a weak focus of sufficiently high order. In contrast, when the focus is strong, by pulling back rays through the origin in the linearized system, we can find families of *isochones* through which the time evolution is constant. Further details can be found in the paper [70].

As with the case of Lyapunov quantities, one can obtain the critical periods as polynomials in the coefficents of the monomoials of (1.4). In analogy to the center-focus problem, we can then ask for the conditions under which a family of systems has an isocronous center.

It can be shown that an isochronous center can be brought to the form (1.11) by a change of coordinates. Rather than bringing the system to the form (1.12), we can therefore compute the conditions for an isochronous center by seeking series

$$X = x + O(2), \qquad Y = y + O(2),$$

which brings the system (1.2) to the system (1.11).

In one sense, seeking for linearizable centers is an easier task computationally. This is because there are effectively double the number of conditions at each degree and the resulting system of equations can often be tractable even when the center-focus problem is not.

On the other hand, the linearizability condition would appear to be slightly less natural, in that there are a number of different mechanisms which can be shown to underlie an isochronous center and for the moment, no attempt has been made to give an equivalent to Żołądek's conjecture for these systems.

Problem 1.1. Develop a list of mechanisms underlying the existence of isochronous centers which covers the known cases.

The concept of isochronicity and, more generally, the study of isochrones appears to have useful applications in the field of mathematical biology and elsewhere (see, for example, the comments in [99]).

1.3 Bifurcation of Limit Cycles from Centers

As mentioned above, for a generic finite family of analytic vector fields, the existence of a center has infinite codimension, and therefore cases of centers will not be expected to appear. But in families of polynomial systems, the set of parameters which give centers form significant strata in the set of all polynomial vector fields. The strata therefore are likely organizing centers for the behavior of the systems in their neighborhood in parameter space.

In this last section, we give a nice application of how the knowledge of a strata of the center variety in a family of systems can give good estimates of the number of limit cycles in the whole family.

If we count free parameters in the expression for the return map $h(c)$ in (1.8), we would expect that in general the codimension of the center variety should be one more than the number of limit cycles that can bifurcate from the center as we move away from the center variety. Though this is not true in general, it does seem to hold in many cases. Furthermore, as we show below, it is often sufficient just to look at the linearization of the Lyapunov quantities to determine this.

Suppose that the coefficients of (1.4) depend polynomially on a finite set of parameters Λ, which includes the parameter λ. We choose a transversal at the origin and calculate the return map $h(c)$ as before. The limit cycles of the system are locally given by the roots of the expression

$$P(c) = h(c) - c = \alpha_1 c + \sum_{i=2}^{\infty} \alpha_k c^k,$$

where the α_i are analytic functions of Λ.

We are interested in a fixed point of the parameter space, K, which we can without loss of generality choose to be the origin (λ must be zero at a bifurcation point, and the other parameters can be translated appropriately).

More detailed calculations show that $\alpha_1 = e^{2\pi\lambda} - 1 = 2\pi\lambda(1 + O(\lambda))$ and that

$$\alpha_k = \beta_k + \sum_{i=1}^{k-1} \beta_i w_{ik}, \quad (k > 1)$$

where the β_i are polynomials in the coefficients of p and q. The w_{ik} are analytic functions of Λ. We set $\beta_1 = 2\pi\lambda$. Furthermore, β_{2k} always lies in the ideal generated by the previous β_i ($1 \leq i \leq 2k - 1$) in the polynomial ring generated by the coefficients in Λ. This means that in the calculations below the β_{2i} turn out to be almost redundant. The β_{2i+1} are of course just the Lyapunov quantities $L(i)$.

Suppose now that at the origin of K, we have $L(i) = 0$ for all i, then the critical point is a center. Let $\mathbb{R}[\Lambda]$ denote the coordinate ring generated by the parameters $\Lambda = \{\lambda_0, \ldots, \lambda_r\}$, with $\lambda_0 = \lambda$, and I the ideal generated in this ring by the Lyapunov quantities. As above, the Hilbert basis theorem shows that there is some number N for which the first N of the $L(i)$ generate I.

Since all the β_{2k}'s lie in the ideal generated by the $L(i)$ with $i < k$, we can write

$$P(c) = \sum_{i=0}^{N} b_{2i+1} c^{2i+1} (1 + \Psi_{2i+1}(c, \lambda_0, \dots, \lambda_r)), \qquad (1.14)$$

where the functions Ψ_{2i+1} are analytic in their arguments and $\Psi(0,0) = 0$. A standard argument from [14] shows that at most N limit cycles can bifurcate.

To find the cyclicity of the whole of the center variety, not only is it necessary to know about the zeros of the $L(i)$, but also the ideal that they generate. It is no surprise therefore that few complete examples are known of center bifurcations [14, 129].

However, if we work about a specific point on the center variety, we can simplify these calculations greatly. Instead of taking the polynomial ring generated by the $L(i)$, we can take the ideal generate by the $L(i)$ in $\mathbb{R}\{\{\Lambda\}\}$, the power series ring of Λ about $0 \in K$ instead. This also has a finite basis, by the equivalent Noetherian properties of power series rings.

What makes this latter approach so useful is that in many cases this ideal will be generated by just the linear terms of the $L(i)$. In which case we have the following theorem.

Theorem 1.5. *Suppose that $s \in K$ is a point on the center variety and that the first k of the $L(i)$ have independent linear parts (with respect to the expansion of $L(i)$ about s); then s lies on a component of the center variety of codimension at least k, and there are bifurcations which produce $k - 1$ limit cycles locally from the center corresponding to the parameter value s.*

If, furthermore, we know that s lies on a component of the center variety of codimension k, then s is a smooth point of the variety, and the cyclicity of the center for the parameter value s is exactly $k - 1$.

In the latter case, $k - 1$ is also the cyclicity of a generic point on this component of the center variety.

Proof. The first statement is obvious. As above we can without loss of generality choose s to be the origin. Since the theorem is local about the origin of K, we can perform a change of coordinates so that the first k of the $L(i)$ are given by λ_i.

Now since we can choose the λ_i independently, we can take $\lambda_i = m_i \epsilon^{2(k-i)}$ for some fixed values m_i $(0 \le i \le k - 1)$, and $m_k = 1$. The return map will therefore be an analytic function of ϵ and c. From (1.14) above, we see that

$$P(c)/c = \sum_{i=0}^{k} m_i c^{2i} \epsilon^{2(k-i)} + \Phi(c, \epsilon).$$

Here Φ contains only terms of order greater than $2k$ in c and ϵ. For appropriate choices of the m_i, the linear factors of $\sum_{i=0}^{r} m_i c^{2i} \epsilon^{2(k-i)}$ can be chosen to be distinct and real, and none tangent to $\epsilon = 0$; whence $P(c)/c$ has an ordinary $2k$-fold point at the origin as an analytic function of c and ϵ. Now it is well known that in this case each of the linear factors $c - v_i \epsilon$ of the terms of degree $2k$ can be extended to an analytic solution branch $c = v_i \epsilon + O(\epsilon^2)$ of $P(c)/c = 0$. This gives $2k$ distinct zeros for small ϵ, and the second statement follows.

The third statement follows from noticing that the first k of the $L(i)$ must form a defining set of equations for the component of the center variety. Any $L(i)$ for $i > k$ must therefore lie in the ideal of the $L(i)$ if we work over $\mathbb{R}\{\{\Lambda\}\}$. The result follows from Bautin's argument mentioned above [14].

The last statement follows from the fact that the points where the center variety is not smooth, or where the linear terms of the first k Lyapunov quantities are dependent, form a closed subset of the component of the center variety we are on. $\qquad\square$

Armed with this result, we can do two things. One is to try to find complete components of the center variety by comparing the dimension of a known algebraic subset of the center variety with its codimension calculated above. Another is to try to find some family of centers of high codimension to see how many limit cycles we can produce. We give two examples of the latter.

Theorem 1.6. *There exists a class of cubic systems with* 11 *limit cycles bifurcating from a critical point. There exists a class of quartic systems with* 15 *limit cycles bifurcating from a critical point.*

Proof. We first consider the family of cubic systems C_{31} in Żołądek's most recent classification [132]. These systems have a Darboux first integral of the form

$$\phi = \frac{(xy^2 + x + 1)^5}{x^3(xy^5 + 5xy^3/2 + 5y^3/2 + 15xy/8 + 15y/4 + a)^2}. \tag{1.15}$$

There is a critical point at

$$x = \frac{6(8a^2 + 25)}{(32a^2 - 75)}, \qquad y = \frac{70a}{(32a^2 - 75)}.$$

If we translate this point to the origin and put $a = 2$ we find we have the system,

$$
\begin{aligned}
\dot{x} &= 10(342 + 53x)(289x - 2112y + 159x^2 - 848xy + 636y^2), \\
\dot{y} &= 605788x - 988380y + 432745xy - 755568y^2 + 89888xy^2 - 168540y^3,
\end{aligned}
$$

whose linear parts give a center.

We consider the general perturbation of this system in the class of cubic vector fields. That is, we take a parameter for each quadratic and cubic term and also a parameter to represent λ above, when the system is brought to the normal form (1.4).

Routine computations now show that the linear parts of $L(0), \ldots, L(11)$ are independent in the parameters and therefore 11 limit cycles can bifurcate from this center.

For the quartic result, we look at a system whose first integral is given by

$$\phi = \frac{(x^5 + 5x^3 + y)^6}{(x^6 + 6x^4 + 6/5xy + 3x^2 + a)^5}. \tag{1.16}$$

The form is inspired by Żołądek's system C_{45} in [132]. We take $a = -8$ which gives a center at $x = 2$, $y = -50$, which we move to the origin. This gives a system

$$\begin{aligned} \dot{x} &= -510x - 6y - 405x^2 - 3xy - 120x^3 - 15x^4, \\ \dot{y} &= 49950x + 510y + 22500x^2 - 1335xy - 15y^2 \\ &\quad +2850x^3 - 630x^2y - 300x^4 - 105x^3y. \end{aligned} \tag{1.17}$$

This time we take a general quartic bifurcation and find that the linear parts of $L(0)$ to $L(15)$ are independent. Hence we can produce 15 limit cycles from this center by bifurcation. □

Remark 1.7. The results in this section can be generalized to take second-order terms in the Lyapunov quantities. If we do so, we find that the quartic system above can actually generate 17 limit cycles.

We give one final result, which uses centers given by both Darboux first integrals and symmetries.

Theorem 1.8. *There exists a quartic system with* 22 *limit cycles. The cycles appear in two nests of 6 cycles and one nest of* 10.

Proof. We work with the cubic center C_{45}, which was the one considered by Żołądek in [131]. This is of the form

$$\dot{x} = 2x^3 + 2xy + 5x + 2a, \dot{y} = -2x^3a + 12x^2y - 6x^2 - 4ax + 8y^2 + 4y, \tag{1.18}$$

with first integral

$$\phi = \frac{(x^4 + 4x^2 + 4y)^5}{(x^5 + 5x^3 + 5xy + 5x/2 + a)^4}. \tag{1.19}$$

When $a = 3$, the system has a center at the point $(-3/2, -11/4)$. We translate the system by $(x, y) \mapsto (x - 1, y + 3)$, which brings the critical point to $(-5/2, 1/4)$. Now we perform a singular transformation $(x, y) \mapsto (x, y^2)$. After multiplying the resulting equation through by y we get the quartic system

$$\begin{aligned} \dot{x} &= y(2x^3 + 6x^2 + 2xy^2 + 5x + 2y^2 + 7), \\ \dot{y} &= -3x^3 + 6x^2y^2 - 30x^2 + 12xy^2 - 57x + 4y^4 - 16y^2. \end{aligned} \tag{1.20}$$

This system has a center at the origin, and we calculate that the linear parts of the Lyapunov quantities $L(0)$ to $L(11)$ are independent.

Now, suppose we add perturbation terms to the system (1.23) in such a way that applying the same transform as that given above we still obtain a system of degree 4. Clearly any perturbation of this form does not affect the center of (1.20) which is given by symmetry.

Furthermore, we can calculate that the new perturbation terms have the linear parts of $L(0)$ to $L(7)$ independent and so can produce 6 limit cycles, which will be doubled by the singular transformation. Thus we have 22 limit cycles in all. □

Notes

§1 The calculation of center conditions for quadratic and homogeneous cubic systems is well known and we do not repeat them here. Detailed accounts can be found in [115, 117]. A similar result for cubic systems is well beyond the computational capabilities of even the most powerful computers. Apart from these "standard" results, there are a very large number of finite families of polynomial systems for which center conditions have been calculated. We do not try to summarize them here. A common e-resource for known center conditions (especially the many families of cubic systems that have been discovered) put in a common format and classified according to type would be a real bonus to further research in this area.

The calculation of Lyapunov quantities for other systems is also well-trodden ground. Algorithms have been implemented in various ways by many authors. Again, we do not attempt to survey them here. Generally, speaking the generation of Lyapunov quantities is usually the straightforward part. The real computational difficulties arise as we try to find their common zeros. The book by Romanovski and Shafer explains these techniques and results in some detail [110].

§2 We have not considered at all the equivalent of the center-focus problem for degenerate centers. That is, degenerate critical points with neighborhoods consisting of closed trajectories. The decision problem for whether a general family of monodromic critical points is a center or not for certain parameter values has been shown to be non-algebraic by Il'yashenko (see the account in [5]). There have also been attempts to apply holonomy techniques in the analytic case to show that such points can have more complicated mechanisms which govern the production of a center [16].

§3 Although our interest is at the moment in real centers, there are good reasons for working over the complex numbers. We can take the existence of a local analytic first integral as the definition of a center in this case. In the case of a real saddle which has a local first integral, we will also use the term *integrable saddle*. We can bring a complex saddle with 1:-1 eigenvalues to the form

$$\dot{x} = x + p(x, y), \qquad \dot{y} = -y + q(x, y), \tag{1.21}$$

which is the complex analog of a weak focus or center. One can calculate Lyapunov quantities for (1.21) exactly as before; these are better known as *saddle quantities* in this case. It seems that the various classes of complex centers arising in quadratic and symmetric cubic systems intersect much more naturally with the real integrable saddles than with the real centers [57] and many authors have considered the integrability problem in this case.

Unfortunately, it is not easy to give a geometric interpretation to integrability for real saddles without resorting to their embedding in the complex numbers. However, in the case of a homoclinic loop attached to a saddle, the saddle quantities contribute distinct asymptotic terms to the return map of the saddle loop [81, 113, 114].

§4 On a similar note, one can consider the more general case of integrable saddles with $p : -q$ resonance; that is, saddle points with ratio of eigenvalues $-p/q$ where the system can be brought after a change of coordinates and time scaling into the form

$$\dot{x} = x, \qquad \dot{y} = -\frac{p}{q}y.$$

As with the isochronous case there are no well-defined lists of mechanisms under which a

Problem 1.2. Develop a sufficient list of mechanisms underlying the existence of integrable $p : -q$ resonant saddles.

Some progress on this problem has been made by Żołądek[133]. In general this is a very hard problem: even in the case of Liénard systems, whose center conditions can be described completely (see Chapter 4), the $p : -q$ case has a very different flavour [72].

§5   There is a very close connection between the study of planar polynomial systems and the theory of holomorphic foliations of codimension 1. We will will only mention one nice application of the center-focus problem to holomorphic foliations here. Suppose ω is an integrable polynomial 1-form in $\mathbb{C}^n$ of degree 2: that is, $d\omega \wedge \omega = 0$. Since ω is integral, about any non-singular point in $\mathbb{C}^n$ we have a local analytic first integral. Restricting to a general 2-plane, we get a quadratic system whose critical points must be integrable. Cerveau and Lins Neto [31] have shown that from the knowledge of the classification of centers of quadratic systems it is possible to classify all the possible forms ω can take.

§6 There is a corresponding Center-Focus problem for higher dimensional systems. For convenience, we will just consider the three dimensional case,

$$\dot{x} = P(x, y, z), \quad \dot{y} = Q(x, y, z), \quad \dot{z} = R(x, y, z). \tag{1.22}$$

Assume that the system, after a change of coordinates and a time scaling, has the form

$$\dot{x} = -y + p(x, y, z), \quad \dot{y} = x + q(x, y, z), \quad \dot{z} = \lambda z + r(x, y, z), \tag{1.23}$$

where p, q and r are terms of degree 2 or more in x, y and z. Clearly the linear system ($p = q = r = 0$) has a center in the plane $z = 0$. When $z \neq 0$, the trajectories either spiral towards or away from this plane, depending on the sign of λ. If we include the nonlinear terms, then we will have a center manifold tangent to $z = 0$ at the origin, with the same pattern of trajectories spiralling towards or away from the manifold. We therefore seek conditions under which the center manifold itself has a center.

It turns out that if we have a center, then the center manifold will be analytic and there will be a first integral of the form $x^2 + y^2 + \cdots$. In other cases, we can calculate a Liapunov function of the form $V = x^2 + y^2 + v_3(x, y, z) + \cdots$ as before with

$$\frac{d}{dt} V = \sum_{i=1}^{\infty} \alpha_i (x^2 + y^2)^i,$$

where the α_i are the Lyapunov quantities in this case. This method was described by [123] and applied in [125] to show that the three dimensional Lotka-Volterra system:

$$\begin{aligned}
\dot{x} &= x\,(\lambda + ax + by + cz), \\
\dot{y} &= y\,(\mu + dx + ey + fz), \\
\dot{z} &= z\,(\nu + 1 + gx + hy + kz),
\end{aligned} \tag{1.24}$$

can have at least 5 limit cycles bifurcating from the (unique) critical point lying outside the coordinate planes.

The calculation of the Lyapunov quantities in the 3D case involve some very large expressions even for lower order quantities, even before we come to solving them. The work [125] uses a clever choice of conditions to try to avoid this but, in general the problem is extremely difficult.

However, the techniques in Section 1.3 can be applied if we work from a known center condition, and reduce the problem to one of linear algebra. In particular, it can be shown that (1.24) with the center manifold given by an invariant plane can bifurcate 2 limit cycles, and when given by an invariant conic can bifurcate 4 limit cycles [116].

It would be nice to have a list of mechanisms which can show integrability in these cases. It would seem that a variety of new techniques will need to be found. In the related case of a $p : -q : r$ resonant saddle at the origin, considered at the end of Section 9, the arguments needed to prove integrability are quite varied. The Darboux method is still valid, replacing invariant curves by invariant sufaces, but the symmetric centers do not form a component since they force the system to have a non-isolated singularity at the origin.

Problem 1.3. Classify the center conditions for the system (1.24) at the critical point not on the coordinate planes.

This would seem to be a very difficult problem, although some progress has been made by Żołądek and Bobieński [17]

§7 More details of the calculations for the center bifurcations can be found in
[39], from which the examples in the last section were drawn.

The calculations in Section 1.3 have been taken up by Torregrosa and others
to give some very strong estimates for the cyclicity of higher degree systems [82, 83,
106]. These works involves an innovative application of parallel computation to give
answers for systems of high degree. Systems with symmetries, as in Theorem 1.8,
have also been tackled in this way in order to give high lower bounds in terms of
the degree d for both the local cyclicity of a center or focus as well as the global
cyclicity (the so-called *Hilbert numbers, $H(d)$*).

A new and interesting feature in [106] is the application of center bifurcation
technique to fine foci with a very high order. Usually, fine foci of high order are
very difficult to compute, but when the system is given by terms with different
symmetries, it is possible to find such points. For example, consider the system

$$\dot{z} = iz + \bar{z}^{n-1} + z^n, \tag{1.25}$$

which gives rise to a system of the form (1.2) when we take $z = x + iy$ and split
into real and imaginary parts. The first and second terms in (1.25) have a $\mathbb{Z}_{n-1}$
symmetry given by $z \mapsto \omega z$, with $\omega^{n-1} = 1$ and the first and third terms a $\mathbb{Z}_n$
symmetry of a similar form. The effect of the both together is a system whose
fine focus can be proved to be of order $(n-1)^2$ for $n \leq 100$. The techniques
of Section 1.3 are then applied to show that the system is indeed a fine focus
and estimate the number of limit cycles that can be obtained by bifurcation. It
would be interesting to know if these examples could be generalised to ones with
more symmetries. Unfortunately, the order of fine focus for (1.25) is not known for
general n. Tools which would allow us to understand this problem better would
be very useful.

Problem 1.4. Show that the system (1.25) has a fine focus of order $(n-1)^2$ for all
values of n. Can systems with, say, three or more competing symmetries be used
to improve the large n cyclicity bounds further?

§8 A growing area of research in recent years is the study of piecewise systems.
These are systems which the plane is divided into two or more regions where
the dynamics are given by different polynomial vector fields. The behaviour of
these systems is very rich, even for linear or quadratic systems. Due to lack of
expertise, we have not attempted to give further details on these systems in this
work. However, the center-focus problem, and the associated center mechanisms
for piecewise systems would make an interesting study.

Additional matieral on integrability and other topics can be found in [128].

Chapter 2

Darboux Integrability

In this chapter, we consider one of the two main mechanisms which seem to underlie the existence of centers in polynomial vector fields. The background and history to this topic is covered in detail by Schlomiuk [117].

2.1 Invariant Algebraic Curves

We consider the system (1.2). For the statements of the following definitions and propositions it is often more convenient to work with the associated vector field (1.1).

Definition 2.1. Let $f \in \mathbb{C}[x, y]$. If the algebraic curve $f = 0$ is invariant by a vector field X of degree d, then $X(f)/f$ is a polynomial of degree at most $d - 1$. In this case we say that $f = 0$ is an *invariant algebraic curve* of X and $L_f = X(f)/f$ is its *cofactor*.

> We remark that many other names, for example *Darboux factor*, have also been used for this concept.

> Note that, if the vector field X has several invariant algebraic curves of different degrees, the cofactors will all lie in $\mathbb{C}_{d-1}[x, y]$, the vector space of polynomials of degree at most $d - 1$. This allows us to reduce the problem of Darboux integrability to one of linear algebra. The proof of the next proposition is clear.

Proposition 2.2. *Let $f \in \mathbb{C}[x, y]$ and $f = f_1^{n_1} \cdots f_r^{n_r}$ be its factorization in irreducible factors. Then, for a vector field X, $f = 0$ is an invariant algebraic curve with cofactor L_f if, and only if, $f_i = 0$ is an invariant algebraic curve for each $i = 1, \ldots, r$ with cofactor L_{f_i}. Moreover $L_f = n_1 L_{f_1} + \cdots + n_r L_{f_r}$.*

Definition 2.3. Let $f, g \in \mathbb{C}[x, y]$; we say that $e = \exp(g/f)$ is an *exponential factor* of the vector field X of degree d, if $X(e)/e$ is a polynomial of degree at most $d - 1$. This polynomial is called the *cofactor* of the exponential factor e, which we denote by L_e. The quotient g/f is an *exponential coefficient* of X.

© The Author(s), under exclusive license to Springer Nature Switzerland AG 2024
C. Christopher et al., *Limit Cycles of Differential Equations*, Advanced Courses
in Mathematics - CRM Barcelona, https://doi.org/10.1007/978-3-030-59656-9_2

Exponential factors represent the coalescence of two or more invariant algebraic curves and so appear natural in families of vector fields with invariant algebraic curves. An example of the phenomena is given in Example 2.14

Proposition 2.4. *If $e = \exp{(g/f)}$ is an exponential factor for the vector field X, then f is an invariant algebraic curve and g satisfies the equation*

$$X(g) = gL_f + fL_e ,\qquad(2.1)$$

where L_f is the cofactor of f.

The existence of invariant algebraic curves has strong consequences for the dynamics of a polynomial system. For example, a quadratic system with an invariant hyperbola or ellipse, cannot have a limit cycle, except for the ellipse itself, and a qudratic system with an invariant parabola or line can have at most one limit cycle.

2.2 The Darboux Method

We are interested in the role of invariant algebraic curves in constructing first integrals and integrating factors of *Darboux* type: that is, functions which are expressible as products of invariant algebraic curves and exponential factors. We recall the following definitions.

Definition 2.5. Let P/Q be a rational function in x and y, with P and Q coprime, then its *degree* is the maximum of the degrees of P and Q.

Definition 2.6. A (multi-valued) function is said to be *Darboux* if it is of the form

$$e^{g/h} \prod_{i=1}^{r} f_i^{l_i},\qquad(2.2)$$

where the f_i, g and h are polynomials, and the l_i are complex numbers.

We shall see in the next chapter that the set of such functions is precisely the set of exponentials of integrals of closed rational 1-forms in x and y.

Definition 2.7. Let U be an open subset of $\mathbb{C}^2$. We say that a non-constant function $H : U \to \mathbb{C}$ is a *first integral* of a vector field X on U if, and only if, $X|_U(H) = 0$. When H is the restriction of a rational (resp. Darboux) function to U, then we say that H is a *rational (resp. Darboux) first integral*.

Definition 2.8. We say that a non-zero function $R : U \to \mathbb{C}$ is an *integrating factor* of a vector field X on U if, and only if, $X(R) = -R\operatorname{div}(X)$ on U, where div denotes the divergence of the vector field.

If we know an integrating factor we can compute, by quadrature, a first integral of the system. Reciprocally, if H is a first integral of the vector field (1.2), then there is a unique integrating factor R satisfying

$$RP = -\frac{\partial H}{\partial y} \quad \text{and} \quad RQ = \frac{\partial H}{\partial x} . \tag{2.3}$$

Such R is called the *integrating factor associated to H*.

In practice, it is usually more natural to work with the reciprocal of an integrating factor, called the *inverse integrating factor*, $1/R$, rather than R itself.

A theorem of Singer [119] shows that if H is a Liouvillian function, then the integrating factor is Darboux. In an earlier work, Prelle and Singer [108] showed that if H is an elementary function, then the integrating factor is the N-th root of a rational function. We shall demonstrate Singer's theorem in the next chapter.

The idea behind the Darboux method is to use the invariant algebraic curves of the system to find an integrating factor of the form (2.2). This, in turn, is purely a matter of linear algebra since, from Proposition 2.2, all the cofactors lie in $\mathbb{C}_{d-1}[x, y]$. A simple introduction to these things can be found in [43].

For example, we can find a Darboux first integral (2.2) if we can find constants l_i and m_i such that

$$\sum_{i=1}^{r} l_i L_{f_i} + \sum_{j=1}^{s} m_j L_{e_j} = 0,$$

where the L_{f_i} and L_{e_j} represent the cofactors of f_i and $\exp(g_j/h_j)$ respectively. In particular, this will always happen if there are more than $d(d+1)/2$ such curves or exponential factors.

Proposition 2.9. *Let X be a vector field. If X admits p distinct invariant algebraic curves $f_i = 0$, for $i = 1, \ldots, p$, and q independent exponential factors e_j, for $j = 1, \ldots, q$. Then the following statements hold.*

(a) *There are $\lambda_i, \rho_j \in \mathbb{C}$, not all zero, such that $\sum_{i=1}^{p} \lambda_i L_{f_i} + \sum_{j=1}^{q} \rho_j L_{e_j} = 0$ if and only if the (multi-valued) function $f_1^{\lambda_1} \cdots f_p^{\lambda_p} e_1^{\rho_1} \cdots e_q^{\rho_q}$ is a first integral of the vector field X.*

(b) *There are $\lambda_i, \rho_j \in \mathbb{C}$, not all zero, such that $\sum_{i=1}^{p} \lambda_i L_{f_i} + \sum_{j=1}^{q} \rho_j L_{e_j} = -\mathrm{div}(X)$ if and only if the function $f_1^{\lambda_1} \cdots f_p^{\lambda_p} e_1^{\rho_1} \cdots e_q^{\rho_q}$ is an integrating factor of X.*

Thus, the problem of finding first integrals or integrating factors is reduced to a question of the linear dependence of the set of cofactors and the divergence of the vector field.

In practice, as stated above, the existence of invariant algebraic curves have very strong consequences on the nature of the system. In most cases it turns out that far less than $d(d+1)/2$ such curves are needed to find a linear dependency for a given system.

When searching for integrability conditions, it is helpful therefore to have a way to reduce the dimension of the space of possible cofactors. In order to do this, we introduce the following concepts from [34].

Proposition 2.10. *Let p be a critical point of the vector field X. Then if f is an invariant algebraic curve of X which does not vanish at p, its cofactor L_f must vanish at p. Furthermore, if $e = \exp(g/f)$ is an exponential factor of X, then L_e must vanish at p too.*

Proof. This follows directly from the equations $X(f) = L_f f$ and $X(g) = L_f g + L_e f$. $\qquad\square$

Definition 2.11. Let X be a vector field of degree d, and $S \subset \mathbb{C}^2$ a finite set of points (possibly empty). The *restricted cofactor space with respect to S*, Σ_S, is defined by

$$\Sigma_S = \cap_{p \in S} m_p \cap \mathbb{C}_{d-1}[x,y] \ ,$$

where m_p is the maximal ideal of $\mathbb{C}[x,y]$ corresponding to the point p.

If S consists of s points, then we say that they are *independent* with respect to $\mathbb{C}_{d-1}[x,y]$ if

$$\sigma := \dim \Sigma_S = \dim \mathbb{C}_{d-1}[x,y] - s = \frac{1}{2}(d+1)(d+2) - s \ .$$

Theorem 2.12. *Let X be a vector field of degree d. Assume that X has p distinct invariant algebraic curves $f_i = 0$, $i = 1,\ldots,p$ and q exponential factors $e_i = \exp(g_i/h_i)$, $i = 1,\ldots,q$, where each h_i is equal to f_k for some k. Suppose, furthermore, that there are s critical points $p_1,\ldots,p_r$ which are independent with respect to $\mathbb{C}_{d-1}[x,y]$, and $f_j(p_k) \neq 0$ for $j = 1,\ldots,p$ and $k = 1,\ldots,r$. Then the following statements hold.*

(a) *If $p + q \geq \sigma + 2$, then X has a rational first integral.*

(b) *If $p + q \geq \sigma + 1$, then X has a Darboux first integral.*

(c) *If $p + q \geq \sigma$, and $\mathrm{div}(X)$ vanishes at the p_i, then X has either a Darboux first integral or a Darboux integrating factor.*

Proof. Statements (b) and (c) follow from counting dimensions and applying Proposition 2.9. One has just to observe that all possible cofactors are contained in Σ_S by Proposition 2.10.

When $p + q \geq \sigma + 2$, we apply (b) to obtain two independent Darboux first integrals, say H_1 and H_2. We can see easily that the integrating factor R_i associated to $\log H_i$ is a rational function. Since the quotient of two integrating factors is a first integral, the statement (a) follows from the independence of H_1 and H_2. $\qquad\square$

Definition 2.13. If (1.2) has a center given by a Darboux first integral or integrating factor, we call it a *Darboux* center.

Unfortunately, given the degree of the system, there is no bound on the degree of the invariant algebraic curves. In fact systems are known with curves of arbitrary degree [44]. This causes problems in trying to give a method to find all Darboux centers which terminates.

Lins Neto and Cerveau [31] have shown that for curves with at most nodal singularities, the degree of the curve exceeds the degree of the system by at most two. similar bound has been given by Carnicer [28] if the system has no dicritical singularities. That is, it is not possible to obtain a star node by a series of blow ups of the critical points. Equivalently, each critical point only has a finite number of invariant analytic branches (possibly with singularities) passing through it.

The following questions would at least show that the systems (1.2) with invariant algebraic curves form an algebraic subset in the set of parameters.

Problem 2.1. Is there is a number $N(d)$ such that if (1.2) has an invariant algebraic curve, then it has an invariant algebraic curve of degree at most $N(d)$?

A similar conjecture would also be useful for applications to the center-focus problem.

Problem 2.2. Is there a number $N(d)$ such that if (1.2) has an invariant algebraic curve of degree greater than $N(d)$, then it has a Darboux first integral or integrating factor?

Of course, in both cases we would prefer to have some concrete way of determining $N(d)$.

A quadratic system with an invariant algebraic curve of degree 12 but not Darboux integrable is given in [52].

2.3 Multiple Curves and Exponential Factors

If we are interested in families of Darboux centers, then we need to be able to understand how the family will change at the points where one or more curves of the system coalesce. In general, this will give rise to exponential factors, as the following simple example makes clear.

Example 2.14. Consider the vector field

$$X = x\frac{\partial}{\partial x} + ((1+\ell)y + x)\frac{\partial}{\partial y}$$

with invariant algebraic curves $x = 0$ and $x + \ell y = 0$. As ℓ tends to zero, then these two curves coalesce. However we can recover an exponential factor, by taking the limit of the Darboux function $((x + \ell y)/x)^{1/\ell}$ which tends to $\exp(y/x)$.

In general, we would hope that an exponential factor $\exp(f/g)$ corresponds to the coalescence of two invariant algebraic curves $f = 0$ and $f + \ell g = 0$ as ℓ tends to zero. That is, we consider $\exp(f/g)$ as

$$\exp(f/g) = \lim_{\ell \to 0} ((f + \ell g)/f)^{1/\ell}.$$

Coalescence of more that two curves will give rise to additional exponential factors and we would like to understand how this multiplicity can be read off from the dynamics.

Although a completely general explanation of this phenomena is not yet known, we give here a summary of the case when $f = 0$ is given by an irreducible polynomial f. The proofs of these results and precise definitions can be found in [51].

Suppose $X = X_\lambda$ depends on a parameter λ. If X_λ has m invariant algebraic curves $f_{\lambda,i} = 0$, $i = 1, \ldots, m$, which converge to the curve $f = 0$ as λ tends to zero, then we say that the curve $f = 0$ of X_0 has *multiplicity m*.

It turns out that a multiple curve of multiplicity m will have associated to it $m - 1$ exponential factors of the form $\exp(g_i/f^i)$ for g_i a polynomial of degree at most $i \deg(f)$.

How do we detect such multiple curves without knowing, a priori, a family which they lie in? It turns out that there are several equivalent definitions of multiplicity of an algebraic curve and a couple of these are very computational in form.

First, we can associate to a multiple curve a *generalized invariant algebraic curve* of the form

$$F = f_0 + \varepsilon f_1 + \cdots + \varepsilon^{k-1} f_{k-1}, \qquad f_0 = f,$$

where each of the polynomials f_i have degree at most $\deg(f)$, and ε is an algebraic quantity with $\varepsilon^k = 0$. That is, we have a curve with some "infinitesimal" information attached. This generalized invariant algebraic curve satisfies

$$X(F) = FL_F, \qquad L_F = L_0 + \varepsilon L_1 + \cdots + \varepsilon^{k-1} L_{k-1},$$

where each of the L_i has degree at most $d - 1$.

Conversely, it is easy to see that if we have such a generalized curve satisfying the above equation, then the coefficients of ε^i of

$$\log(F) = \log(f_0) + \varepsilon \frac{f_1}{f_0} + \varepsilon^2 \frac{f_2 - f_1^2/2}{f_0^2} + \cdots = \log(f_0) + \varepsilon g_1 + \varepsilon^2 g_2 + \cdots$$

give $k - 1$ exponential factors, e^{g_i/f^i}, with $\deg(g_i) \leq i \deg(f_0)$.

A more computational approach to finding this multiplicity is given by computing the extactic:

$$
\mathcal{E}_n = \det \begin{pmatrix} v_1 & v_2 & \cdots & v_l \\ X(v_1) & X(v_2) & \cdots & X(v_l) \\ \vdots & \vdots & \cdots & \vdots \\ X^{l-1}(v_1) & X^{l-1}(v_2) & \cdots & X^{l-1}(v_l) \end{pmatrix}, \tag{2.4}
$$

where $n = \deg(f)$, and $v_1, v_2, \ldots, v_l$ is a basis of $\mathbb{C}_n[x, y]$, the $\mathbb{C}$-vector space of polynomials in $\mathbb{C}[x, y]$ of degree at most n, with $l = (n+1)(n+2)/2$, and where we define $X^0(v_i) = v_i$ and $X^j(v_i) = X^{j-1}(X(v_i))$. It turns out that the multiplicity can also be given by the maximum power of f appearing in $\mathcal{E}_n$.

Finally, computations of multiplicity can also be obtained from the holonomy group, which we examine in Chapter 9. Here we also need to impose some restrictions on the critical points which can appear on $f = 0$. Further details for all these equivalent definitions can be found in [51].

Problem 2.3. Generalize the notion of algebraic curve to allow for more degerate bifurcations (i.e. families of conics appearing from two lines or from a double line).

Notes

§1 Darboux functions are ubiquitous in the area of polynomial systems. For example, when the system is non-integrable, but has sufficient limit cycles, they can be used to give geometric non-existence results for limit cycles [43]. Another area of application is for showing not only integrability but linearizability [93]. Here we look for substitutions of the form

$$
X = x\, m(x, y), \qquad Y = y\, n(x, y),
$$

where m and n are Darboux functions, in order to bring the system to a linear form.

§2 Movasati [97] has shown that given integers d_i, $i = 1, \ldots, r$ with $\sum d_i = d+1$, the set of centers which have a Darboux first integral

$$
\prod_{i=1}^{r} f_i^{l_i}, \qquad \deg(f_i) = d_i
$$

form a complete component of the center variety for systems of degree d. The idea is to consider higher order perturbations from an associated Hamiltonian system. A similar technique has been used by Zare and Tanabé to show that systems with generic symmetries form irreducible components of the center variety [127].

§3 The notion of multiplicity allows one to naturally obtain complete families of systems with a certain number of invariant algebraic curves. For example, cubic systems with a fixed number of invariant lines have been investigated in [26, 120].

§4 Given a set of curves, $f_i = 0$ of degree d_i, $i = 1 \cdots r$, it would be nice to know the general form of a system for which these curves are invariant, the so-called *inverse problem*. When the union of the curves with the line at infinity form a normal crossing divisor (that is, at most two curves cross at any point and that only transversally), then it is possible to show that the system must take the form

$$\dot{x} = A K - \sum_{i=1}^{r} D_i K_{iy}, \qquad \dot{y} = B K + \sum_{i=1}^{r} D_i K_{ix}, \tag{2.5}$$

where A and B are polynomials of degrees $d - \sum d_i$, and the D_i have degree $d + 1 - \sum d_i$. In particular, the sum of the d_i cannot be greater than $d + 1$. When the sum is equal to $d + 1$, then A and B vanish and the D_i must be constants, forcing the system to have a Darboux first integral $\prod f_i^{D_i}$.

Similar formulas to (2.5) can be found for systems with exponential factors, and more generally for systems with higher multiple curves [51]. In the case of just one curve, the dimension of the systems not satisfying (2.5) can be given in terms of the singularities of the curve [47].

§5 Much of the theory here can be carried across to systems of higher dimension, replacing invariant algebraic curves with invariant hypersufaces. A multiplicity can be defined in the same way for such surfaces [86]. Darboux first integrals can then be constructed in a similar way as Theorem 2.12.

One significant difference, however, is that the generalisation of an integrating factor to the higher dimensional cases. We call a function satisfying the equation $X(R) = -R \operatorname{div}(X)$ a *Jacobi multiplier*. Unlike the two dimensional case, the knowledge of a Jacobi multiplier does not imply the existence of a first integral. For a system of dimension n, however, knowing $n - 2$ first integrals and a Jacobi multiplier will, in favourable circumstances, allow the construction of another first integral and hence show that the system is completely integrable.

For three dimensional systems, in particular, the restriction of a Jacobi multiplier to a level surface of a first integral gives an integrating factor on the surface, and hence a first integral can be constructed on the level surface. By choosing these first integrals in a consistent way, the first integral can be defined more globally.

A good resource for Jacobi multipliers is [15].

§6 I'd like to mention one problem here which is quite intriguing though of minor importance. Given a system (1.2) with a Darboux integrating factor, what can be said about its first integral. In the case where the invariant algebraic curves $f_i = 0$ are in generic position, and we have no exponential factors, we can also find a Darboux first integral.

This has been shown algebraically in [45], however it also has a nice geometric interpretation. Let D be the Darboux integrating factor, then

$$\phi = \int D(P\,dy - Q\,dx)$$

defines a multi-valued integral outside the set Z of zeros of the curves $\{f_i = 0\}$. The effect of passing around a non-trivial loop γ in the complement of Z takes ϕ to $h_\gamma(\phi) = a_\gamma \phi + b_\gamma$ for some constants a_γ and b_γ. If the set Z together with the line at infinity has only nodal singularities (which will be true in the generic case when the $f_i = 0$ are smooth and intersect transversally with each other and infinity), it is well known that the complement of Z has an abelian fundamental group. This means that the maps h_γ must commute. It is then straightforward to show that either $a_\gamma = 1$ for all γ or the addition of a constant to ϕ makes all the b_γ vanish. In the first case, $\phi - \sum c_i \log(f_i)$ is single-valued for some constants c_i given by the b_γ, and in the second, $\phi / \prod f_i^{m_i}$ is single-valued for some constants m_i given by the a_γ. Application of growth estimates shows that these functions are therefore rational and hence gives either $\exp(\phi)$ or ϕ as a Darboux first integral. In the case when D contains exponential terms, then it is necessary to use the extended monodromy group of Żołądek [134].

In general, it seems that the first integrals have the property that they are a sum of a Darboux function plus a sum of several one-dimensional integrals of the form

$$\int^{h(x,y)} e^{s(u)} \prod r_i(u)_i^\lambda,$$

where h, s and the r_i are rational functions. Żołądek calls these *Darboux–Schwartz–Christoffel* integrals, and conjectures that all first integrals of Darboux centers can be obtained in this form. Interesting examples can be found in [134].

Chapter 3

Liouvillian Integrability

In this chapter we want to prove that Darboux integrability corresponds to the notion of Liouvillian integrability, or "solution by quadratures".

3.1 Differential Fields and Liouvillian Extensions

Let (K, Δ) be a *differential field*. That is, a field K equipped with a set of commuting operators $\delta : K \to K$, $\delta \in \Delta$, called *derivations* satisfying

$$\delta(x + y) = \delta x + \delta y, \qquad \delta(xy) = (\delta x)y + x(\delta y).$$

We shall assume that all fields have characteristic zero.

A *differential field extension* $(K', \Delta') \supset (K, \Delta)$ is a field extension $K' \supset K$ for which the restriction of each $\delta' \in \Delta'$ to K is given by some $\delta \in \Delta$. Because of this compatibility condition we can use Δ to represent both sets of derivations without confusion.

From now on we will drop the explicit references to the derivations Δ of the differential fields unless necessary.

Example 3.1. If K' is an algebraic extension of a differential field K, then the derivations of K extend to K' in an unique way. This does not hold true if the characteristic of the field is different from zero.

Example 3.2. If K is a differential field, we can form the extension field $K' = K(t)$, for some t transcendental over K. To say that this is a differential extension is to say that there are elements $a_\delta \in K(t)$ such that $\delta t = a_\delta$ for all $\delta \in \Delta$. The commutativity of the derivations means that $\delta_1 a_{\delta_2} = \delta_2 a_{\delta_1}$ for all $\delta_1, \delta_2 \in \Delta$.

Definition 3.3. A differential field extension $K \supset k$ is called *Liouvillian* if it can be written as a tower of differential extensions

$$k = K_0 \subset K_1 \subset \cdots \subset K_n = K,$$

© The Author(s), under exclusive license to Springer Nature Switzerland AG 2024
C. Christopher et al., *Limit Cycles of Differential Equations*, Advanced Courses
in Mathematics - CRM Barcelona, https://doi.org/10.1007/978-3-030-59656-9_3

where at each step we have one of the following conditions:

(i) K_{i+1} is a finite algebraic extension of K_i.

(ii) $K_{i+1} = K_i(t)$ for some t with $\delta t/t \in K_i$ for all $\delta \in \Delta$.

(iii) $K_{i+1} = K_i(t)$ for some t with $\delta t \in K_i$ for all $\delta \in \Delta$.

That is, our new field contains functions which can be got by successive (i) solutions of algebraic equations, (ii) exponentials of integrals and (iii) integrals.

We also require the the field of *constants* of K, that is those elements $k \in K$ with $\delta k = 0$ for all $\delta \in \Delta$, is the the same as the field of constants of K_0. This makes sure that we do not add spurious first integrals when extending K

In our applications we will always take the extensions starting from the field $K_0 = \mathbb{C}(x, y)$ with the standard derivations $\Delta = \{\partial/\partial x, \partial/\partial y\}$. We shall write $\partial f/\partial x$ for $(\partial/\partial x)f$.

We say that (1.2) has a Liouvillian first integral if there exists a function ϕ in some Liouvillian extension of $\mathbb{C}(x, y)$ such that $X(\phi) = 0$, where X is the associated vector field in (1.1), and at least one of $\partial\phi/\partial x$ and $\partial\phi/\partial y$ is not zero.

Theorem 3.4 (Singer [119]). *The system* (1.2) *has a Liouvillian first integral, if and only if it has a Darboux integrating factor.*

In fact, Singer's original proof [119] shows slightly more than this. He proves that if a polynomial system has a trajectory whose phase curve is given by the zeros of a Liouvillian function, then the trajectory is either algebraic, or we have a Liouvillian first integral. This demonstrates something of the importance of invariant algebraic curves in the analysis of polynomial systems.

The "if" statement in the theorem is clear: if D is our Darboux integrating factor, then $\delta D/D$ is a rational function and we can therefore add D to our field by one extension of type (ii). The converse will be proved in the next section.

3.2   Proof of Singer's Theorem

Proposition 3.5. *If the system* (1.2) *has a Liouvillian first integral, then it has an integrating factor of the form*

$$\exp\left(\int U\,dx + V\,dy\right), \qquad \frac{\partial U}{\partial y} = \frac{\partial V}{\partial x},$$

where U and V are rational functions.

Proof. From the hypothesis of the theorem, there exists a ϕ in some Liouvillian extension field K of $\mathbb{C}(x, y)$ with $X(\phi) = 0$. Thus,

$$h\frac{\partial\phi}{\partial x} = Q, \qquad h\frac{\partial\phi}{\partial y} = -P,$$

for some $h \in K$. Thus, putting

$$A = \frac{1}{h}\frac{\partial h}{\partial x}, \qquad B = \frac{1}{h}\frac{\partial h}{\partial y}$$

we have elements A and B in K such that

$$PA + QB = \frac{\partial P}{\partial x} + \frac{\partial Q}{\partial y}, \qquad \frac{\partial A}{\partial y} = \frac{\partial B}{\partial x}. \tag{3.1}$$

We want to show that if the above equation can be satisfied for A and B in K_{i+1} it can be satisfied in K_i. The conclusion follows directly from the case $i = 0$, putting $U = A$ and $V = B$.

We consider each type of extension in turn. Without loss of generality, we can assume that the extensions in (ii) and (iii) are transcendental, else we consider them under (i).

(i) Let $\tilde{K}_{i+1}$ be the normal closure of K_{i+1}. Let Σ be the set of automorphisms of K_{i+1} fixing K_i. We also write N for $|\Sigma|$. Then,

$$\sum_{\sigma \in \Sigma} \sigma(PA + QB) = N\Big(\frac{\partial P}{\partial x} + \frac{\partial Q}{\partial y}\Big),$$

and

$$\frac{\partial \sigma(A)}{\partial y} = \frac{\partial \sigma(B)}{\partial x},$$

for all σ in Σ. Hence, taking

$$\bar{A} = \frac{1}{N}\sum_{\sigma \in \Sigma}\sigma(A), \qquad \bar{B} = \frac{1}{N}\sum_{\sigma \in \Sigma}\sigma(B),$$

we have

$$P\bar{A} + Q\bar{B} = \frac{\partial P}{\partial x} + \frac{\partial Q}{\partial y}, \qquad \frac{\partial \bar{A}}{\partial y} = \frac{\partial \bar{B}}{\partial x},$$

where $\bar{A}$ and $\bar{B}$ must lie in K_i.

(ii) We have $A = a(t)$ and $B = a(t)$ for some rational functions a and b with coefficients in K_i. Since t is transcendental, we can expand $a(t)$ and $b(t)$ formally as Laurent series in t. Let a_0 and b_0 be the coefficients of t^0 in the expansions of $a(t)$ and $b(t)$ respectively, then the coefficient of t^0 in (3.1) can be seen to give

$$Pa_0 + Qb_0 = \frac{\partial P}{\partial x} + \frac{\partial Q}{\partial y}, \qquad \frac{\partial a_0}{\partial y} = \frac{\partial b_0}{\partial x}.$$

(iii) We take $A = a(t)$ and $B = b(t)$ as above, but now expand formally as a Laurent series in $1/t$. Let r be the highest power of t appearing in the expansion of either $a(t)$ or $b(t)$, and let a_r and b_r be the respective coefficients of t^r in these expansions. Equating powers of t^r in (3.1), we get

$$Pa_r + Qb_r = 0, \qquad \frac{\partial a_r}{\partial y} = \frac{\partial b_r}{\partial x},$$

for $r \neq 0$, and

$$Pa_0 + Qb_0 = \frac{\partial P}{\partial x} + \frac{\partial Q}{\partial y}, \qquad \frac{\partial a_0}{\partial y} = \frac{\partial b_0}{\partial x},$$

for $r = 0$. In the latter case the inductive step is complete, and in the former we can find an element $h \in K_i$ such that

$$P = -b_r h, \qquad Q = a_r h.$$

Then

$$\frac{\partial P}{\partial x} + \frac{\partial Q}{\partial y} = P\frac{1}{h}\frac{\partial h}{\partial x} + Q\frac{1}{h}\frac{\partial h}{\partial y},$$

and taking

$$A = \frac{1}{h}\frac{\partial h}{\partial x}, \qquad B = \frac{1}{h}\frac{\partial h}{\partial y},$$

we complete the inductive step in this case also.

Thus we can repeat the inductive step until we can find solutions to (3.1) in $K_0 = \mathbb{C}(x, y)$ and we are finished.   □

To finish the proof of Theorem 3.4 we need the following proposition.

Proposition 3.6. *If the system* (1.2) *has an integrating factor of the form*

$$\exp\left(\int U\,dx + V\,dy\right), \qquad \frac{\partial U}{\partial y} = \frac{\partial V}{\partial x},$$

where U and V are rational functions of x and y, then there exists an integrating factor of the form

$$\exp(g/f)\prod f_i^{l_i},$$

where g, f and the f_i are polynomials in x and y.

Proof. Let K be a normal algebraic extension of $\mathbb{C}(y)$ which is a splitting field for the numerators and denominators of U and V considered as polynomials in x over $\mathbb{C}(y)$. We can thus rewrite U and V in their partial fraction expansions

$$U = \sum_{i=1}^{r}\sum_{j=1}^{n_i}\frac{\alpha_{i,j}}{(x - \beta_i)^j} + \sum_{i=0}^{N}\gamma_i x^i, \qquad V = \sum_{i=1}^{\bar{r}}\sum_{j=1}^{\bar{n}_i}\frac{\bar{\alpha}_{i,j}}{(x - \bar{\beta}_i)^j} + \sum_{i=0}^{\bar{N}}\bar{\gamma}_i x^i,$$

where the $\alpha_{i,j}$, $\bar{\alpha}_{i,j}$, β_i, γ_i and $\bar{\gamma}_i$ are elements of K. By taking $\alpha_{i,j}$, $\bar{\alpha}_{i,j}$, γ_i and $\bar{\gamma}_i$ to be zero outside their defined values, we can neglect the explicit mention of the summation limits without confusion.

We now apply the condition $U_y = V_x$ to the above expressions. Gathering terms and using the uniqueness of the partial fraction expansion, we see that (using f' to denote df/dx)

$$\gamma_i' = \bar{\gamma}_{i+1}(i+1), \qquad \alpha_{i,j+1}' + j\beta_i'\alpha_{i,j} + j\bar{\alpha}_{i,j} = 0. \tag{3.2}$$

In particular we have $\alpha_{i,1}' = 0$.

We now write down a function and show that it is indeed the integral $\int U\,dx + V\,dy$. Let ϕ be given by

$$\phi = \sum \alpha_{i,1} \log(x - \beta_i) + \sum \frac{\alpha_{i,j}}{(x - \beta_i)^{j-1}}\left(\frac{-1}{j-1}\right) + \sum \frac{\gamma_i x^{i+1}}{i+1} + \int \bar{\gamma}_0\,dy,$$

where the last term represents any primitive of $\bar{\gamma}$. It is easy to verify that $\partial\phi/\partial x = U$ and $\partial\phi/\partial y = V$ using (3.2).

We now let Σ represent the group of automorphisms of K over $\mathbb{C}(y)$, and let $N = |\Sigma|$. As in the previous proof, we define

$$\bar{\phi} = \frac{1}{N}\sum_{\sigma \in \Sigma} \sigma(\phi),$$

where we note that

$$\sigma(\alpha_{i,1}\log(x - \beta_i)) = \alpha_{i,1}\log(x - \sigma(\beta_i)),$$

and

$$\sigma\left(\int \bar{\gamma}_0\,dy\right) = \int \sigma(\bar{\gamma}_0)\,dy,$$

is only defined up to an arbitrary constant.

It is clear that we still have $\partial\bar{\phi}/\partial x = U$ and $\partial\bar{\phi}/\partial y = V$. Furthermore we have

$$\bar{\phi} = \sum l_i \log(R_i(x,y)) + R(x,y) + \int S(y)\,dy,$$

where R_i, R and S are rational functions. We can evaluate the integral in the last term via the partial fraction expansion of S as

$$\int S(y)\,dy = \sum \alpha_i \log(S_i(y)) + S_0(y),$$

where the S_i are polynomials in y. Taking exponentials, the integrating factor obtained is of the form desired. $\qquad\square$

3.3   Riccati equations

We mention briefly here another possible mechanism which guarantees that a critical point is a center. However, although we shall give non-trivial examples of this mechanism for the integrability of a saddle, we do not know of any non-trivial example of such a case for real centers. These systems exhibit both properties of symmetric systems in that there is a reduction to a simpler system, and Darboux systems, in that there is a first integral in the form of a Darboux first integral, but where the factors are solutions of a second-order differential equation.

We first give an example of a $1 : -\lambda$ resonant saddle which is integrable via the solutions of a Riccati equation.

Example 3.7. The system

$$\dot{x} = x(1-x), \qquad \dot{y} = -\lambda y + d_1 x^2 + d_2 xy + y^2, \tag{3.3}$$

with $\lambda > 0$ but not an integer has a first integral

$$\phi = x^{\lambda} \frac{(y - \alpha x)F_1(x) + x(1-x)F_1'(x)}{(\lambda - y + (\alpha - \lambda)x)F_2(x) - x(1-x)F_2'(x)}, \tag{3.4}$$

where $F_1(x) = F(a, b; c; x)$ and $F_2(x) = F(a - c + 1, b - c + 1; 2 - c; x)$ where $F(a, b; c; x)$ is the Gauss hypergeometric function

$$F(a, b; c; x) = \sum_{n=0}^{\infty} \frac{(a)_n (b)_n}{(c)_n n!} x^n$$

with

$$(a)_n = \begin{cases} a(a+1)\ldots(a+n-1) & n \geq 1 \\ 1 & n = 0 \end{cases}$$

and similarly for $(b)_n$ and $(c)_n$. Here we define α as a root of $\alpha^2 - (\lambda - d_2)\alpha + d_1 = 0$, $c = \lambda + 1$ and a and b are the roots of $A^2 - (1 + 2\alpha + d_2)A + \alpha(\lambda + 1) = 0$.

It is easy to see that a transformation $x \mapsto X^n$ will give non-trivial examples of integrable $1 : -1$ saddles. The functions in the numerator and denominator of (3.4) satisfy a similar equation to the polynomials defining invariant algebraic curves. In particular, they have polynomial cofactors.

The following example appears quite naturally in the class of cubic systems with a $1 : -1$ saddle (see [109]).

Example 3.8. Consider the system

$$\dot{x} = x - 9bx^3 - ay^3, \qquad \dot{y} = -y + bxy - 6b^2 x^2 y. \tag{3.5}$$

We perform a transformation

$$X = y^3(1 - 3bx)^{-2}, \qquad Y = x - aX/4,$$

which brings the system to the form

$$\dot{X} = -3X - 6abX^2, \qquad \dot{Y} = Y + \frac{39}{16}a^2bX^2 + \frac{9}{2}abXY + 3bY^2,$$

which can be brought to the form of Example 3.7 by a simple scaling of the X and Y axes.

It would be interesting to consider extending the work in the previous chapter to include solutions of second-order linear differential equations to see if the above types of integrals are the only ones.

Notes

§1 There are several simplified proofs of Singer's theorem in the literature [29, 103, 133]. We have followed loosely the one in Pereira [103]. Singer's original proof is in [119]. The proof can be carried over very simply to the case of holomorphic foliations of codimension 1, and has been done so by several of the authors cited.

If we restrict items (ii) and (iii) in Theorem 3.4 to just the addition of exponentials and logarithms,

(ii)′ $K_{i+1} = K_i(t)$ for some t with $\delta t/t = g$ for some $g \in K_i$ and all $\delta \in \Delta$,

(iii)′ $K_{i+1} = K_i(t)$ for some t with $\delta t = \delta g/g$ for some $g \in K_i$ and all $\delta \in \Delta$,

we obtain the concept of an *elementary* first integral. This corresponds to integrals in "closed form" (without quadratures). In an earlier paper, Prelle and Singer [108] showed that all elementary first integrals have integrating factors which are fractional powers of a rational function in x and y. The proof is similar to the Liouvillian case but makes use of Rosenlicht's result, Theorem 5.1. There have been attempts to implement these results as symbolic integration routines. One indication of how this can be done is given in [91]. Liouvillian integrability is discussed in [6]. There are also some partial results known about Liouvillian integrability for higher order differential equations [7]. In [13], a detailed examination of Liouvillian first integrals of three dimensional systems is given.

§2 The role of Riccati equations for the center problem was discussed by Żołądek in [133], who also gave the example analyzed in Example 3.7. These systems have quite a distinct behaviour and have been used to give interesting constructions, for example a family of quadratic systems with invariant algebraic curves of arbitrary degree [44] mentioned in the previous chapter.

§3 Of course it would also be interesting to examine other types of extensions. Casale [29] gives results for extensions involving the solution of Riccati equations. The question is closely tied with the existence of Godbillon–Vey sequences of finite length. Those of length 1 correspond to Darboux integrable systems, and those of length 2 to Riccati systems. It is not clear whether higher length sequences play a role in planar systems or not. For some suggestive results in this direction see [32, 33].

§4 Another problem of interest is to describe all systems which have Elementary or Liouvillian solutions. When the invariant curves of the system together with the line at infinity form a normal crossing divisor then it is possible to describe all such systems [47]. This has been generalised in [48, 49]. Some results on the inverse problem for systems with elementary first integrals can be found in [50]. However, as mentioned at the end of the last chapter, the general case of the form of first integral for a system with a Darboux integrating factor, or even (in the elementary case) an integrating factor which is the root of a rational function, is still unknown.

Chapter 4

Symmetry

In this chapter we consider the second mechanism which gives rise to centers in polynomial systems: the existence of an algebraic symmetry.

After some brief preliminary comments, we shall show that such symmetries can be used to obtain a complete classification of centers in polynomial Liénard systems.

4.1   Algebraic Symmetries

Let $(x, y) \mapsto (X(x,y), Y(x,y))$ be an analytic transformation which is a local involution in the neighborhood of the origin. After a linear transformation we can assume that

$$X = x + r(x, y), \qquad Y = -y + s(x, y),$$

where s is quadratic or higher in x and y. Choosing new coordinates

$$\xi = \frac{x + X(x,y)}{2} = x + O(2), \qquad \zeta = \frac{y - Y(x,y)}{2} = y + O(2),$$

the involution is brought to the form $(\xi, \zeta) \mapsto (\xi, -\zeta)$ and so we have a symmetry with respect to the line $\zeta = 0$.

If the critical point at the origin is monodromic, then it is clear that the symmetry condition implies that we have a center. However, we can see this in another way which will also apply to the case of complex centers.

In the new coordinates, (1.2) becomes

$$\dot{\xi} = \zeta \tilde{P}(\xi, \zeta^2) h(\xi, \zeta), \qquad \dot{\zeta} = \tilde{Q}(\xi, \zeta^2) h(\xi, \zeta), \qquad (4.1)$$

with $h(0,0) \neq 0$. We can remove the factor h without loss of generality, as we are only interested in the orbital behavior of the system. Doing this, and taking $Z = \zeta^2$, we obtain (after again ignoring a common factor ζ) the *reduced* system

$$\dot{\xi} = \tilde{P}(\xi, Z), \qquad \dot{Z} = 2\tilde{Q}(\xi, Z). \qquad (4.2)$$

© The Author(s), under exclusive license to Springer Nature Switzerland AG 2024
C. Christopher et al., *Limit Cycles of Differential Equations*, Advanced Courses
in Mathematics - CRM Barcelona, https://doi.org/10.1007/978-3-030-59656-9_4

We call the map $(\xi, \zeta) \mapsto (\xi, \zeta^2)$ a *reducing transformation*. Now, (4.2) no longer has a singular point at the origin, and hence there is a local first integral $\phi(\xi, Z)$ in the neighborhood of the origin. The pull back of this first integral via the reducing transformation gives a first integral $\phi(\xi, \zeta^2)$ of (4.1), and hence (4.1) has a center at the origin (see Figure 4.1).

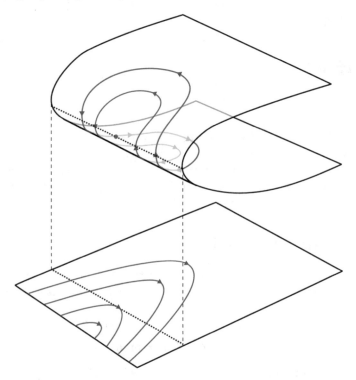

Figure 4.1: Pulling back along a fold line to obtain a center.

We saw in Chapter 1 that every center is orbitally equivalent to the linear center by an analytic change of coordinates, and hence every center has an analytic symmetry. So the existence of an analytic symmetry does not carry with it much information about the real "cause" of the center.

To understand the mechanisms behind a center-focus problem, we will only be interested in functions $X(x, y)$ and $Y(x, y)$ which are algebraic over $\mathbb{C}(x, y)$. This gives us a nice algebraic and global mechanism for a center.

Example 4.1. The Kukles' system

$$\dot{x} = y, \qquad \dot{y} = -x + a_1 x^2 + a_2 xy + a_3 y^2 + a_4 x^3 + a_5 x^2 y + a_6 xy^2 + a_7 y^3,$$

has a symmetry in the x-axis for $a_2 = a_5 = a_7 = 0$, and a symmetry in the y-axis for $a_1 = a_3 = a_5 = a_7 = 0$. From the symmetry in the x-axis we obtain a reducing transformation $Z = y^2$, and a reduced system

$$\dot{x} = 1, \qquad \dot{Z} = 2(-x + a_1 x^2 + a_3 Z + a_4 x^3 + a_6 x Z),$$

having removed a common factor y from the system.

Those centers which are given by algebraic symmetries seem to form components of the center variety of relatively large dimension compared with the Darboux centers. In the case of systems of low degree, however, the reduced systems tend to be sufficiently simple that they can be algebraically integrated, and so are Darboux as well as symmetric.

We want to show that algebraic symmetries comprise all the centers for Polynomial Liénard systems. In order to do this, we need a nice characterization of analytic conditions for a center, given by Cherkas.

4.2   Centers for analytic Liénard equations

In this section we describe the analytic conditions for a center for the system

$$\dot{x} = y, \qquad \dot{y} = -g(x) - y f(x), \tag{4.3}$$

where f and g are real polynomials in x. This was first obtained by Cherkas in [35].

The system (4.3) arises from the second-order nonlinear equation

$$\ddot{x} + f(x)\dot{x} + g(x) = 0, \tag{4.4}$$

which generalizes the van der Pol oscillator, and is ubiquitous in the study of polynomial systems.

Any critical point of (4.3) lies on the x-axis. We can assume, after a translation of the x-axis, that the critical point we are interested in is at the origin. The condition that this critical point should be non-degenerate and a focus or center implies that $g(0) = 0$ and $g'(0) > 0$. It also implies that $f(0)^2 < 4g'(0)$, however we do not need this condition here.

We now wish to transform (4.3) into a more amenable form. Denote

$$F(x) = \int_0^x f(\xi)\, d\xi, \qquad G(x) = \int_0^x g(\xi)\, d\xi.$$

Under the Liénard transformation $y \mapsto y + F(x)$, the system is brought to the form

$$\dot{x} = y - F(x), \qquad \dot{y} = -g(x). \tag{4.5}$$

We can simplify (4.5) further by a transformation which effectively removes g. Let u be the positive root of $2G$. From the conditions on g given above it is clear that this root is well defined and analytic in a neighborhood of $x = 0$, Thus

$$u = (2G(x))^{1/2} \operatorname{sgn}(x) = (g'(0))^{1/2} x + O(x^2) \tag{4.6}$$

defines an invertible analytic transformation in a neighborhood of $x = 0$.

Let $x(u)$ denote the inverse of this function. The transformation (4.6) takes the system (4.5) to the system

$$\dot{u} = \frac{g(x(u))}{u}\left(y - F(x(u))\right), \qquad \dot{y} = -g(x(u)).$$

Since $g(x(u))/u = (g'(0))^{1/2} + O(u)$ is analytic and non-zero in a neighborhood of the origin, we can rescale (4.6) by multiplying the right-hand side by $u/g(x(u))$ which gives

$$\dot{u} = y - F(x(u)), \qquad \dot{y} = -u. \tag{4.7}$$

This system has exactly the same direction field as (4.6) in a neighborhood of the origin, and hence the local qualitative behavior of the system, in particular the existence of a center, is not altered by this scaling.

We write the power series for $F(x(u))$ as $\sum_1^\infty a_i u^i$. It turns out that the origin of (4.7) is a center if and only if all the a_{2i+1} vanish. To see this we introduce the function $F^*(u) = \sum_1^\infty a_{2i} u^{2i}$, analytic in a neighborhood of the origin, and consider the system

$$\dot{u} = y - F^*(u), \qquad \dot{y} = -u. \tag{4.8}$$

Since the flow of (4.8) is monodromic, it is clear that it must have a center at the origin due to symmetry in the u-axis. However, the system (4.7) is rotated with respect to (4.8) in a neighborhood of the origin unless all the terms a_{2i+1} vanish.

Thus (4.7) cannot have a center at the origin unless all the a_{2i+1} vanish. On the other hand if all the a_{2i+1} do vanish, then the system is a center by symmetry.

We can express this necessary and sufficient condition in a more geometrical form:

Theorem 4.2. *The system* (4.3) *has a center at the origin if and only if* $F(x) = \Phi(G(x))$, *for some analytic function* Φ, *with* $\Phi(0) = 0$.

Proof. The argument above shows that there is a center if and only if $F(x(u)) = \phi(u^2)$ for some analytic function ϕ, $\phi(0) = 0$. But $u^2 = 2G(x)$, so set $\Phi(w) = \phi(2w)$. $\qquad\square$

Now consider the function $z(x)$ defined in a neighborhood of the origin by $z(x) = x(-u(x))$. We can also describe $z(x)$ as the unique analytic function which satisfies

$$G(x) = G(z), \qquad (z(0) = 0, \ z'(0) < 0).$$

That this equation defines a unique analytic function $z(x)$ is clear from the conditions on g since

$$G(x) - G(z) = (x - z)\left(\tfrac{1}{2}g'(0)(x + z) + o(x, z)\right) = 0$$

has two analytic branches at the origin $z = x$ and $z = -x + o(x)$. The conditions on $z'(0)$ then selects the second of these. Now $2G(x(u)) = u^2 = 2G(x(-u))$ whence

$G(x) = G(x(-u(x)))$. Furthermore, as $x(-u(x)) = -x + O(x^2)$, this solution must be $x(-u(x))$.

We know that the origin is a center if and only if the function $F(x(u))$ is even. That is, $F(x(u)) - F(x(-u))$ vanishes identically. But this is equivalent to saying that $F(x(u(x))) - F(x(-u(x))) = F(x) - F(z) = 0$. Thus, we have the following characterization of centers.

Theorem 4.3. *The system* (4.3) *has a center at the origin if and only if there exists a function* $z(x)$ *satisfying*

$$F(x) = F(z), \qquad G(x) = G(z), \qquad (z(0) = 0, \ z'(0) < 0). \qquad (4.9)$$

This result also works when f and g are only analytic functions in a neighborhood of the origin. However, if f and g are also polynomials, then the solution $z(x)$ must correspond to a common factor between the functions $F(x) - F(z)$ and $G(x) - G(z)$ other than $(x - z)$. Thus, the following corollary is clear:

Corollary 4.4. *If the system* (4.3) *with* f *and* g *polynomials has a center at the origin, then it is necessary that the resultant of*

$$\tilde{F} = \frac{F(x) - F(z)}{x - z} \quad \text{and} \quad \tilde{G} = \frac{G(x) - G(z)}{x - z} \qquad (4.10)$$

with respect to x *or* z *vanishes. This condition is sufficient if the common factor of the two polynomials vanishes at* $x = z = 0$.

In fact, it can be shown that the order of intersection of $\tilde{F}$ and $\tilde{G}$ at the origin is twice the order of weak focus. Although simple to compute for low degree systems [53], the asymptotic behaviour of this intersection number is still poorly understood.

Problem 4.1. Determine the maximum order of intersection of $\tilde{F}$ and $\tilde{G}$ in (4.10). How far does this number differ from the cyclicity of the origin?

The latter question is relevant as it is shown in [53] that the cyclicity for complex systems can be strictly larger.

4.3 Centers for polynomial Liénard equations

Corollary 4.4 gives algebraic conditions for a center, but does not indicate how systems satisfying these conditions arise. We now want to show that the centers are in fact given by algebraic symmetries in the x variable.

Consider the subfield of $\mathbb{R}(x)$ generated by the polynomials F and G. Call this field $\mathcal{F}$. The field $\mathcal{F}$ shares an important property with F and G:

Lemma 4.5. *Suppose there exists an analytic function* $z(x)$ *with* $z(0) = 0$, $z'(0) < 0$ *such that both* $F(z(x)) = F(x)$ *and* $G(z(x)) = G(x)$ *in a neighborhood of* $x = 0$, *Then for all elements* H *of the field* $\mathcal{F}$ *generated by* F *and* G *we have* $H(z(x)) = H(x)$ *considered as meromorphic functions of* x *about* $x = 0$.

Proof. Note first that $H(z(x)) = 0$ if and only if $H(x) = 0$. Thus we need only verify that addition, multiplication and inversion of non-zero elements of $\mathcal{F}$ preserve this property, which is clearly the case. $\qquad\square$

Recall that Lüroth's Theorem states that if k is a field, any subfield of $k(x)$ which strictly contains k is isomorphic to $k(x)$. That is to say, the subfield is just $k(r)$ for some $r \in k(x)$. But $\mathcal{F}$ is a subfield of $\mathbb{R}(x)$ strictly containing $\mathbb{R}$, and so we must have $\mathcal{F} = \mathbb{R}(r)$ for some rational function $r \in \mathbb{R}(x)$.

Let us write r as A/B with $A, B \in \mathbb{R}[x]$. The field generated over $\mathbb{R}$ by r can also be generated by

$$\frac{\alpha r + \beta}{\gamma r + \delta} = \frac{\alpha A + \beta B}{\gamma A + \delta B}$$

for any constants α, β, γ and δ, such that $\alpha\delta - \beta\gamma \neq 0$. By choosing these constants to ensure that the degree of the denominator is less than the degree of the numerator, we can assume without loss of generality that B has degree less than A. We can also assume that all common factors between A and B are canceled and that B is monic.

Now F and G are in $\mathcal{F}$, and so

$$F = \frac{F_1(A, B)}{F_2(A, B)}, \qquad G = \frac{G_1(A, B)}{G_2(A, B)},$$

where the F_i and G_i are homogeneous polynomials which we choose to have no common factors as polynomials in A and B. We now show that:

Lemma 4.6. $B = 1$.

Proof. We first factor the expressions for F_1 and F_2 over $\mathbb{C}[A, B]$ to obtain

$$F_1(A, B) = \prod_{i=1}^{r}(\lambda_1 A + \mu_i B), \quad F_2(A, B) = \prod_{i=r+1}^{r+s}(\lambda_1 A + \mu_i B),$$

for some complex constants λ_i and μ_i.

Now note that if $\lambda_1 A + \mu_1 B$ and $\lambda_2 A + \mu_2 B$ have a common factor as polynomials in x, then they are multiples of each other, since A and B have no common factors in x (over $\mathbb{R}$ and therefore over $\mathbb{C}$ too). However we chose F_1 and F_2 to have no common factors as polynomials in A and B, hence they must have no common factors as polynomials in x.

Thus the denominator of F as a rational function of x after cancellation with the numerator is just $F_2(A(x), B(x))$, and so

$$\prod_{i=r+1}^{r+s}(\lambda_1 A + \mu_i B) \in \mathbb{R}.$$

Since the degree of A is larger than the degree of B, this can only happen when $\lambda_i = 0$ for all $i = r + 1, \ldots, s$ and hence B must be a constant polynomial, and therefore equal to 1. Similar considerations show that $G_2(A(x), B(x))$ is also a constant. □

Thus we have shown that both F and G are polynomials of some polynomial $A \in \mathcal{F}$. The final step follows.

Theorem 4.7. *The system* (4.3) *with* $g(0) = 0$ *and* $g'(0) > 0$ *has a non-degenerate center at the origin if and only if* $F(x)$ *and* $G(x)$ *are both polynomials of a polynomial* $A(x)$ *with* $A'(0) = 0$ *and* $A''(0) \neq 0$.

Proof. By Theorem 4.3, if there is a center at the origin of (4.3), then there is a function $z(x)$ with $z(0) = 0$ and $z'(0) < 0$ such that $F(z(x)) = F(x)$ and $G(z(x)) = G(x)$. By Lemma 4.5, the polynomial generator of $\mathcal{F}$, A also satisfies $A(z(x)) = A(x)$, and hence its linear term must vanish. Now $G(x)$ is a polynomial in A with $G''(0) > 0$, which means that the quadratic term of A cannot vanish.

Conversely, assume F and G are polynomials of a polynomial A with a non-zero quadratic term but no linear term. From the conditions on A, we can find an analytic function satisfying $A(z(x)) = A(x)$ with $z(0) = 0$, $z'(0) < 0$. Clearly F and G must then satisfy condition (4.9) of Theorem 4.3, and the origin is therefore a center. □

Corollary 4.8. *The system* (4.3) *has a non-degenerate center at the point* $x = p$ *if and only if* $g(p) = 0$, $g'(p) > 0$ *and* F *and* G *are both polynomials of a polynomial* A *which satisfies* $A'(p) = 0$ *with* $A''(p) \neq 0$.

Proof. If we shift the x-axis to bring $x = p$ to the origin, then it is clear that the new F and G calculated will differ from the original ones only by a constant. The rest follows quite easily from Theorem 4.7. □

Theorem 4.9. *If the system* (4.3) *has a non-degenerate center at the origin, then the transformation* $x \mapsto z(x)$, *given from the conditions of Theorem 4.3, takes the direction field of* (4.3) *into itself, reversing the directions. Thus the origin has a generalized symmetry.*

Alternatively, the system (4.3) *can be seen to have a center via the reducing transformation* $w \mapsto h(x)$ *for some polynomial* $h(x) = x^2 + O(x^3)$ *from the system*

$$\dot{w} = y, \quad \dot{y} = -m(w) - l(w)y, \tag{4.11}$$

after a scaling. Here l *and* m *are polynomials* $m(0) > 0$, *and the transformation* $v \mapsto h(x)$ *takes a non-critical point at the origin of* (4.11) *and "unfolds" it into the center of* (4.3).

Proof. The first assertion is a direct calculation from condition (6) of Theorem 4.3. The generalized symmetry condition means that trajectories lying in $x \geq 0$ can be mapped onto trajectories in $x \leq 0$, with the points on $x = 0$ being fixed. If we know that the flow encircles the origin, then trajectories sufficiently close to the origin must be closed. Thus if the critical point is known to be of focal type, this generalized symmetry is enough to imply the existence of a center.

For the second part, we take the polynomial A of Theorem 4.7, and consider

$$h(x) = 2\frac{A(x) - A(0)}{A''(0)} = x^2 + O(x^3).$$

Clearly F and G are also polynomials of this polynomial, so that $F = L(h(x))$ and $G = M(h(x))$ for some polynomials L and M. From the condition on $g'(0)$, we see that $M'(0) > 0$. Take $l = L'$ and $m = M'$, then system (4.11) transforms (after scaling by $h'(x)$) to

$$\dot{x} = y, \quad \dot{y} = -h'(x)m(h(x)) - h'(x)l(h(x))y = -g(x) - f(x)y.$$

The origin of (4.11) is not a critical point, but locally the trajectories are of the form

$$w = \alpha - \frac{1}{2m(\alpha)}y^2 + O(y^3)$$

for small values of α, where the $O(y^3)$ term is analytic in α as well as y. The transformation takes these trajectories to the curves

$$x^2 + O(x^3) = \alpha - \frac{1}{2m(\alpha)}y^2 + O(y^3),$$

for α sufficiently small. These trajectories are thus closed curves approximating the ellipses $x^2 + y^2/(2m(0)) = \alpha$, and the origin is a center.   □

Notes

§1 The role of symmetries in proving the existence of centers is of course standard, but a systematic investigation of algebraic symmetries was carried out by Żołądek [130], who coined the name "rationally reversible" for this phenomenon. We have preferred to consider the more general concept of algebraic symmetries, as these seem to be needed in some of the more complex center examples [58].

§2 The classification of centers in Liénard systems can be found in [37]. The argument here can be extended to Liénard systems with integrable $1 : -1$ saddles as well as Liénard systems with more degenerate singularities. In [40] it is shown how a similar argument can characterise isochronous Liénard systems in a similar way to Theorem 4.3.

Theorem 4.10. *The Lienard system* (4.3) *has an isochronous center at the origin if and only if there is an analytic function* $y = y(x)$ *with* $y(0) = 0$ *and* $y'(0) < 0$ *such that* $F(x) = F(y)$ *and*

$$(x - y)^3\, g(x)\, f(y) = (f(y) - f(x))\Big(\frac{(x - y)^4}{4} + (\tilde{F}(y) - \tilde{F}(x))^2\Big), \qquad F(x) = F(y),$$

where $\tilde{F}$ is an indefinite integral of F.

For systems up to degree 34, it can be shown that the only Liénard systems satisfying this condition have $f(x)$ odd and

$$g(x) = x + \frac{1}{x^3} \int_0^x \xi f(\xi) \, d\xi. \qquad (4.12)$$

but no general result is known. For the companion case of a linearizable $1 : -1$ saddles, at least one other family of solutions appears.

Problem 4.2. Show that all polynomial isochronous Liénard systems are of the form above. In the case of $1 : -1$ linearizable saddles, determine what extra conditions can appear.

The case of $p : -q$ resonant saddles in Liénard systems is much harder [70] and appear to require quite different techniques.

§3 Although it may seem that Liénard systems are quite special, it can be shown [121, 89] that any elementary singular point can be brought to Liénard form (with f and g analytic) by an analytic change of variables. The geometric proof of Loray is one of the gems of the subject.

§4 In the case of integrable saddles we can still apply the symmetry method. However, an interesting phenomena occurs when considering a family of symmetric critical points. Suppose we have a family of saddles depending on a parameter β,

$$\dot{x} = x + p(x, y), \qquad \dot{y} = -y + q(x, y),$$

with the symmetry, $(x, y, -t) \mapsto (\beta y, \beta^{-1} x, -t)$. This implies that

$$p(x, y) = -\beta \, q(\beta y, \beta^{-1} x), \qquad q(x, y) = -\beta^{-1} p(\beta y, \beta^{-1} x);$$

the second of these equations being equivalent to the first. If we write

$$p = \sum_{i, j \geq 0} a_{ij}(\beta) x^i y^j, \qquad q = \sum_{i, j \geq 0} b_{ij}(\beta) x^i y^j,$$

we find that

$$a_{ij}(\beta) = \beta^{1+j-i} b_{ij}(\beta).$$

We can extend the family to include the case $\beta = 0$ if we assume

$$a_{ij}(0) = 0 \ (1 + j > i), \quad a_{ij}(0) + b_{ij}(0) = 0 \ (1 + j = i), \quad b_{ij}(0) = 0 \ (1 + j < i).$$

In this case the system, for $\beta = 0$, can be written as

$$\dot{x} = x(1 + f(xy) + x\tilde{p}(x, xy)), \qquad \dot{y} = y(-1 - f(xy) + y\tilde{q}(x, xy)),$$

for some polynomials $\tilde{p}$, $\tilde{q}$ and f, with $f(0) = 0$. Taking $V = xy$ as a new variable, we get

$$\dot{x} = x(1 + f(V) + x\tilde{p}(x, V)), \qquad \dot{V} = xV(\tilde{p}(x, V) + \tilde{q}(x, V)),$$

and the line of singularities, $x = 0$ can be removed to get a system with no singularities at the origin. The first integral of this system can be pulled back to a first integral of the original system. Effectively, we have blown down to a simpler system for which we know the existence of a first integral.

In the $p : -q$ case this method appears quite often. Here, the blow down is to a node which, once we have shown that any resonant terms vanish, must be integrable and the first integral can be pulled back to the original system. More generally, transformations of the form

$$(X, Y) = (xy^r, y^r),$$

are used for an appropriate choice of rational number r. Examples can be found in [56]. The method occurs frequently in the study of $p : -q : r$ resonant singularities, where the method can be generalized to seek blow down type transformations to a critical point in the Poincaré domain.

Chapter 5

Cherkas' Systems

In this chapter we give an extended example of a non-trivial classification of centers which involves both Darboux and symmetry mechanisms for producing a center. Further details can be found in [58], which we follow closely.

We first need a result concerning algebraic solutions to transcendental equations due to Rosenlicht [112].

Theorem 5.1. *Let* $(k,')$ *be a differential field of characteristic zero with differential extension field* $(K,')$ *with the same field of constants and such that* k *is algebraically closed in* K*, i.e. all elements in* K *which are algebraic over* k *also lie in* k*. We also assume that* K *is a finite algebraic extension of* $k(t)$*, where* t *is transcendental over* k *and such that* $t' \in k$*. Suppose that*

$$\sum_{i=1}^{n} c_i \frac{u_i'}{u_i} + v' \in k,$$

where $c_1, \ldots c_n$ *are constants of* k *which are linearly independent over* $\mathbb{Q}$ *and* $u_1, \ldots u_n$ *and* v *are in* K*. Then* $u_1, \ldots, u_n \in k$ *and* $v = ct + d$*, with* c *a constant of* k *and* $d \in k$*.*

In our applications, we shall take $k = \mathbb{C}$ to ensure that k is algebraically closed. In particular, all elements of k are constants. The application we need of this result could alternatively be obtained by an application of complex variables, but we use the above result to emphasize the algebraic nature of the computations.

We consider the system

$$\dot{x} = y, \qquad \dot{y} = P_0(x) + P_1(x)y + P_2(x)y^2, \tag{5.1}$$

where the $P_i(x)$ are polynomials in x.

We prove the following theorem for this system.

C. Christopher et al., *Limit Cycles of Differential Equations*, Advanced Courses
in Mathematics - CRM Barcelona, https://doi.org/10.1007/978-3-030-59656-9_5

Theorem 5.2. *A system of the form* (5.1) *with* $P_0(0) = 0$ *and* $P_0'(0) < 0$, *which has a center at the origin, satisfies one of the following (possibly overlapping) conditions.*

(i) *The system is algebraically reducible via the map* $(x, y) \mapsto (x, y^2)$ *and thus it has a symmetry in the x-axis.*

(ii) *The system is algebraically reversible at the origin. In fact, it is algebraically reducible via a map* $(x, y) \mapsto (r(x), y)$ *for some polynomial* $r(x)$ *over* $\mathbb{R}$.

(iii) *There is a local first integral of Darboux type.*

Proof. We break the proof into several steps.

Step 1. We perform the change of variables used by Cherkas,

$$y = Y \exp\left(\int_0^x P_2(\xi)\,d\xi\right)$$

to arrive, after renaming the variable Y as y, at the system

$$\dot{x} = y, \qquad \dot{y} = g(x) + f(x)y.$$

Here

$$g(x) = P_0(x)\exp\left(-2\int_0^x P_2(\xi)\,d\xi\right), \qquad f(x) = P_1(x)\exp\left(-\int_0^x P_2(\xi)\,d\xi\right). \quad (5.2)$$

We note that the transformation of Cherkas has changed the system into one which is still polynomial in y but with coefficients in a Liouvillian differential field extension of $(\mathbb{C}(x), d/dx)$ generated by adjoining the exponentials of integrals in (5.2). On the other hand the above transformation has reduced the system to a Liénard form, making it possible to apply the results of Cherkas (Theorem 4.2).

Step 2. We now apply Cherkas' Theorem.

It is easy to verify that the conditions on $P_0(x)$ given above imply the hypotheses of Theorem 4.3 on $g(x)$. Therefore the conclusion of Theorem 4.3 tells us that (5.1) has a center at the origin if and only if there is a real analytic function $z(x)$ in the neighborhood of the origin, with $z(0) = 0$ and $z'(0) = -1$ which simultaneously satisfies

$$\int_0^x f(\xi)\,d\xi = \int_0^z f(\xi)\,d\xi, \qquad \int_0^x g(\xi)\,d\xi = \int_0^z g(\xi)\,d\xi, \quad (5.3)$$

or equivalently,

$$f(x)\,dx = f(z)\,dz, \qquad g(x)\,dx = g(z)\,dz. \quad (5.4)$$

We first dismiss the trivial case where $f(x)$ vanishes identically, as this implies that P_1 is identically zero. The origin is, in this case, a center by symmetry in the

x axis. This is just Condition (i) of the theorem. Alternatively, the system can be algebraically reduced to the system

$$\dot{\bar{x}} = 1, \qquad \dot{\bar{y}} = 2P_0(\bar{x}) + 2P_2(\bar{x})\bar{y}$$

by the map $(x, y) \mapsto (\bar{x}, \bar{y}) = (x, y^2)$. This means that the first integral defined at the origin of this system in $(\bar{x}, \bar{y})$ can be pulled back to a first integral in a neighborhood of the origin of (5.1), giving a center.

Thus we shall assume from now on that f and g do not vanish identically. We shall also exclude the case where P_2 vanishes identically in (5.1) as this has been covered in the previous chapter. In fact, this case is just a subcase of Case 1 below.

From (5.4) we obtain

$$g(x)/f(x) = g(z)/f(z), \tag{5.5}$$

as local meromorphic functions in x around the origin (in the right side of (5.5), we write z for $z(x)$ to simplify notation) and hence $z(x)$ satisfies the equation

$$\frac{P_0(x)}{P_1(x)} \exp\left(-\int_0^x P_2(\xi)\, d\xi\right) = \frac{P_0(z)}{P_1(z)} \exp\left(-\int_0^z P_2(\xi)\, d\xi\right). \tag{5.6}$$

As a real local analytic function, $z(x)$ may be considered as an element of $\mathbb{R}\{\{x\}\}$ which determines an element of $\mathbb{C}\{\{x\}\}$ and hence a local complex analytic function which we also denote by $z(x)$. We now divide our investigation into two cases: the first case is when $z(x)$ is algebraic over $\mathbb{C}(x)$, and the second when $z(x)$ is transcendental over $\mathbb{C}(x)$.

Step 3. $z(x)$ is algebraic over $\mathbb{C}(x)$. In this case we can apply the results of Theorem 5.1. First note that $\int P_2(x)\, dx$ is a non-constant polynomial, and that the equation (5.6) gives

$$\frac{P_0(x)/P_1(x)}{P_0(z)/P_1(z)} \exp\left(-\int_0^x P_2(\xi)\, d\xi + \int_0^z P_2(\xi)\, d\xi\right) = 1. \tag{5.7}$$

Thus we have

$$\Psi(x) = R_1(x, z(x))e^{R_2(x, z(x))} = 1, \tag{5.8}$$

where

$$R_1(x, y) = \frac{P_0(x)/P_1(x)}{P_0(y)/P_1(y)} \in \mathbb{R}(x, y),$$

$$R_2(x, y) = \int_0^y P_2(\xi)\, d\xi - \int_0^x P_2(\xi)\, d\xi \in \mathbb{R}[x, y].$$

$\overline{R_1}(x) = R_1(x, z(x))$ and $\overline{R_2}(x) = R_2(x, z(x))$ therefore lie in the algebraic differential field extension $(\mathbb{C}(x)[z(x)], d/dx)$ of $(\mathbb{C}(x), d/dx)$ generated by $z(x)$.

Below we denote the derivations in the two fields by $'$. Considering now the expression for Ψ'/Ψ from (5.8) we obtain:

$$(\overline{R_1})'/\overline{R_1} + (\overline{R_2})' = 0. \tag{5.9}$$

Thus, by Theorem 5.1, where we take $k = \mathbb{C}$ and $K = \mathbb{C}(x)[z]$, we get that $R_1(x, z(x))$ is a constant and $R_2(x, z(x))$ is of the form $cx + d$, for constants c and d. Substituting back into (5.9), we see that c must vanish. Furthermore, it is clear that at $x = 0$, $R_2(x, z(x))$ vanishes, and hence $d = 0$. Lastly, from (5.8), $R_1(x, z(x)) = 1$.

Thus, we have arrived at the equations

$$P_0(x)/P_1(x) = P_0(z)/P_1(z), \qquad \int_0^x P_2(\xi)\, d\xi = \int_0^z P_2(\xi)\, d\xi. \tag{5.10}$$

Consider the subfield F of $\mathbb{R}(x)$ generated by all rational functions $S(x)$ such that $S(x) = S(z(x))$. By Lüroth's theorem, the field is isomorphic to $\mathbb{R}(r/s)$, for some function $r(x)/s(x)$ with $r(x), s(x) \in \mathbb{R}[x, y]$. Without loss of generality, we can choose the degree of r to be greater than the degree of s, with r and s coprime. Hence, we can write

$$\int_0^x P_2(\xi)\, d\xi = \phi(r(x)/s(x)), \tag{5.11}$$

for some rational function ϕ over $\mathbb{R}$, in one variable.

Now, working over $\mathbb{C}$, the right-hand side of (5.11) can be written as

$$\prod_{i=1}^q (\alpha_i r + \beta_i s) \Big/ \prod_{i=1}^q (\gamma_i r + \delta_i s),$$

As in Section 4.3, if $(\alpha_i r + \beta_i s)$ shares a common factor with $(\gamma_j r + \delta_j s)$, these two polynomials must differ by a constant, whence we can assume that the fraction above allows no further cancellations. Since the left-hand side of (5.11) is a polynomial, $\prod(\gamma_i r + \delta_i s)$ must be a constant, and hence the denominator has no dependence on r, and s must be a constant.

Without loss of generality, we can take $s(x) = 1$ and $r(0) = 0$. Furthermore, from (5.11), we now see that ϕ must be a polynomial.

From (5.10), we must also have

$$P_0(x)/P_1(x) = \psi_0(r/s)/\psi_1(r/s) = \psi_0(r)/\psi_1(r)$$

for some polynomials in one variable ψ_0 and ψ_1 over $\mathbb{R}$ with $(\psi_0, \psi_1) = 1$. From the equality above we get

$$P_0(x)/\psi_0(r(x)) = P_1(x)/\psi_1(r(x)) = K(x),$$

with $K(x)$ a rational function in x over $\mathbb{R}$. Thus,

$$P_0(x) = K(x)\psi_0(r(x)), \qquad P_1(x) = K(x)\psi_1(r(x)), \tag{5.12}$$

which implies that $K(x)$ is a polynomial over $\mathbb{R}$.

Using the expression for P_1 in (5.12) and replacing it into the second part of (5.2) we have:

$$f(x) = K(x)\psi_1(r(x)) \exp\left(-\int_0^x P_2(\xi)\,d\xi\right).$$

Substituting the above in the first part of (5.4) we obtain:

$$K(x)\psi_1(r(x)) \exp\left(-\int_0^x P_2(\xi)\,d\xi\right) dx = K(z)\psi_1(r(z)) \exp\left(-\int_0^z P_2(\xi)\,d\xi\right) dz.$$

But $r(x) = r(z(x))$ and, using the second part of (5.10), we obtain:

$$K(x)\,dx = K(z)\,dz.$$

However, from $r(x) = r(z)$, we also have $r'(x)\,dx = r'(z)\,dz$, and hence,

$$K(x)/r'(x) = K(z)/r'(z).$$

So $K(z)/r'(z)$ is in the field F. This implies that $K(x) = r'(x)\chi(r(x))$ for some rational function χ over $\mathbb{R}$ which, in view of the preceding equality, must be a polynomial by a comparison of the degrees of r and r'. Since $r(x) = r(z)$ with $z'(0) = -1$, we must have $r'(0) = 0$. However, from the expression of $P_0(x)$ in (5.12) and using the expression $K(x) = r'(x)\chi(r(x))$ we get that $P_0'(0) = r''(0)\chi(0)\psi(0)$ and hence $r''(0) \neq 0$. Without loss of generality, we can choose r so that $r''(0) = 1$.

Putting together this information, there exist polynomials A_0, A_1 and A_2 such that

$$P_0(x) = A_0(r(x))r'(x), \qquad P_1(x) = A_1(r(x))r'(x), \qquad P_2(r(x)) = A_2(r(x))r'(x),$$

with $A_i = \chi\psi_i$ for $i = 0, 1$ and $A_2 = \phi'$. The system is then algebraically reducible. Indeed the map $(x, y) \mapsto (\bar{x}, \bar{y}) = (r(x), y)$ reduces (5.1) to the system

$$\dot{\bar{x}} = \bar{y}, \qquad \dot{\bar{y}} = A_0(\bar{x}) + A_1(\bar{x})\bar{y} + A_2(\bar{x})\bar{y}^2.$$

This system is non-singular at the origin since the conditions on P_0 imply that $A_0(0) < 0$.

Alternatively, the center can be seen to be given by a reversing transformation

$$(x, y, t) \mapsto (\bar{x}, y, -t),$$

where $r(x) = r(\bar{x})$, $\bar{x}(0) = 0$ and $\bar{x}'(0) < 0$. This is Condition (ii) in the theorem. We have seen in the previous chapter, that this is also the case when P_2 vanishes identically.

Step 4. z(x) is transcendental over $\mathbb{C}(x)$. In this case, we first consider some consequences of (5.5). Differentiating (5.5) we obtain

$$(g/f)'(x)\, dx = (g/f)'(z)\, dz$$

where $'$ represents the derivative. This gives:

$$\left[\frac{1}{f}\left(\frac{g}{f}\right)'\right](x) = \left[\frac{1}{f}\left(\frac{g}{f}\right)'\right](z),$$

which gives

$$\left[\frac{P_2}{P_1}\left(\frac{P_0}{P_1}\right) - \frac{1}{P_1}\left(\frac{P_0}{P_1}\right)'\right](x) = \left[\frac{P_2}{P_1}\left(\frac{P_0}{P_1}\right) - \frac{1}{P_1}\left(\frac{P_0}{P_1}\right)'\right](z).$$

Since this is an algebraic equation between z and x, and since both x and z are transcendental over $\mathbb{C}$, then both sides of the above equality must be a constant c. In particular, considering the partial fraction expansion of P_0/P_1, we see that P_0/P_1 must, in fact, be a polynomial.

Hence we consider the equality in $\mathbb{R}(x)$:

$$P_2 P_0 P_1 + P_0 P_1' - P_1 P_0' = c P_1^3. \tag{5.13}$$

For $k \in \mathbb{C}$ we define

$$C_k = y + k(P_0/P_1).$$

We seek invariant algebraic curves in the family of curves $C_k = 0$. Recall that a curve $C_k = 0$ is an invariant algebraic curve of the differential system (5.1) if and only if $DC_k/C_k \in C[x,y]$, where D is the operator

$$y\frac{\partial}{\partial x} + (P_0 + P_1 y + P_2 y^2)\frac{\partial}{\partial y}.$$

We determine the condition on the constant k such that this be satisfied.

We first compute

$$DC_k = k(P_0/P_1)'y + P_0 + P_1 y + P_2 y^2.$$

Using (5.13) we have that

$$(P_0/P_1)' = (P_0'P_1 - P_0P_1')/P_1^2 = P_2 P_0/P_1 - cP_1,$$

and hence

$$DC_k = P_2 y^2 + k(P_0 P_2/P_1 - cP_1)y + P_0 + P_1 y. \tag{5.14}$$

We search for polynomials $A_k(x)$ and $B_k(x)$ such that

$$DC_k = C_k(A_k y + B_k) = (y + kP_0/P_1)(A_k y + B_k). \tag{5.15}$$

Clearly, we have from conditions (5.14) and (5.15) that

$$A_k = P_2, \qquad B_k = P_1/k,$$

and

$$-ckP_1 + P_1 = P_1/k,$$

yielding

$$ck^2 - k + 1 = 0.$$

Depending on the value of c this equation has one or two distinct solutions. If $c = 1/4$ it has only one solution $k = 2$, and if $c \neq 1/4$ it has two distinct solutions k_1 and k_2. Thus

$$DC_k = C_k(P_2 y + P_1/k)$$

where $k = 2$ in case $c = 1/4$ and $k = k_j$, $j = 1, 2$ in case $c \neq 1/4$. Note that the k_i are always non-zero.

Before considering each one of these two cases we observe that the system (5.1) admits the expression

$$C = \exp\left(\int_0^x P_2(x)\, dx\right)$$

as an exponential factor, i.e., $DC/C \in \mathbb{C}[x, y]$. Indeed, we have

$$DC = C\,(P_2 y).$$

We now consider the two possible cases. First let us suppose $c \neq 1/4$. In this case we construct a Darboux first integral from these three functions C, C_{k_1} and C_{k_2} of the form

$$\left(y + k_1(P_0/P_1)\right)^{r_1} \left(y + k_2(P_0/P_1)\right)^{r_2} \exp\left(r_3 \int_0^x P_2(x)\, dx\right).$$

It is immediately verified that, if we take $r_1 = 1$, $r_2 = -k_2/k_1$ and $r_3 = -1 + k_2/k_1$, we have the linear combination of their corresponding cofactors

$$r_1(P_2 y + P_1/k_1) + r_2(P_2 y + P_1/k_2) + r_3 P_2 y = 0,$$

which shows that the Darboux function is indeed a first integral.

In the case $c = 1/4$ we only have one invariant algebraic curve, i.e. $C_2 = 0$. We recall that we also have the exponential factor C. We now consider the expression

$$\tilde{C} = \exp(P_0/(P_1 y + 2P_0)),$$

then $\tilde{C}$ is another Darboux exponential factor. Indeed calculations yield

$$D\tilde{C} = \tilde{C}(-P_1/4).$$

In this case we construct a Darboux first integral by using the curve $C_2 = y + 2(P_0/P1)(x)$ and the two exponential factors C and $\tilde{C}$ defined above. This first integral is of the form

$$(C_2)^{r_1} C^{r_2} (\tilde{C})^{r_3}.$$

It is easy to see that if we take $r_1 = 2$, $r_2 = -1$ and $r_3 = 1$ we obtain the following linear combination of their corresponding cofactors:

$$r_1(-P_1/4) + r_2(P_2 y) + r_3(P_2 y + P_1/2) = 0.$$

These integrals are well defined and holomorphic in a neighborhood of the origin, and hence, by Poincaré's result the origin is a center.   □

Notes

Further details can be found in the paper [58] where the more general case

$$\dot{x} = yP(x), \qquad \dot{y} = P_0(x) + P_1(x)y + P_2(x)y^2,$$

with $P(x)$ is a polynomial with $P(0) \neq 0$, is also treated.

We note that in this latter case, the reducing transformations are no longer polynomial, but only algebraic.

There is a method of also dealing with generalized Kukles' systems

$$\dot{x} = yP_0(x), \qquad \dot{y} = P_0(x) + P_1(x)y + P_2(x)y^2 + P_3(x)y^3, \qquad (5.16)$$

where all the P_i are polynomials. The reduction stage is more delicate, but the results are essentially the same: either there is a Darboux first integral, or there is an algebraic symmetry.

In this case, a transformation of the form

$$Y = y \left(1 + y\frac{P_1}{3P_0}\right)^{-1},$$

removes the term in y in (5.16), leaving a similar equation with just terms in Y^0, Y^2 and Y^3. The transformation

$$\tilde{y} = Y \exp(h(x)),$$

for some $h(x)$, removes the term in Y^2, leaving an equation of the form,

$$\dot{x} = \tilde{y}, \qquad \dot{\tilde{y}} = -g(x) - f(x)\tilde{y}^3.$$

This system can be tackled, in much the same way as the Liénard system in the previous chapter, to show that the existence of a center depends on showing that there exists an analytic function, $z(x)$, with $z(0) = 0$ and $z'(0) < 0$, solving the equation

$$F(x) = F(z) \qquad G(x) = G(z),$$

where the functions F and G are given in terms of the original coefficients. Further manipulation of this identity allows one to reduce the problem, as for the Cherkas case, to either an algebraic or transcendental form. These, in turn, show that the existence of a center arises for either an algebraic symmetry or from a Darboux integral [41].

Unfortunately, there does not seem to be any method for tackling more complex systems in this way.

Problem 5.1. Find a differential algebraic reduction of the system

$$\dot{x} = y, \qquad \dot{y} = P_0(x) + P_1(x)y + P_2(x)y^2 + P_3(x)y^3 + P_4(x)y^4,$$

to a generalised Liénard system,

$$\dot{x} = y, \qquad \dot{y} = -g(x) - f(x)y^n,$$

for some n, where the f and g lie in some differential algebraic extension of $\mathbb{C}(x)$. Alternatively, prove that such an extension is impossible.

The transformation could involve solutions of ordinary differential equations, but must be sufficiently explicit so that the type of center mechanisms involved can be deduced by algebraic means. In fact, in Cherkas' original paper on Kukles-type systems, a transformation involving a particular solution, $\phi(x)$, of the equation itself was used to remove the term in y^3.

Chapter 6

Monodromy

In this chapter we begin the second part of these notes, looking at some ideas based around the concept of *monodromy*. Very roughly, this is the study of how objects depending on a parameter, and which are locally constant in some sense, change as the parameter moves around a non-trivial path. If this path is a closed loop, the object often undergoes a non-trivial transformation when the parameter returns to its original value. Understanding these transformations can give great insight into the original problem.

This idea is particularly appropriate for the center-focus problem, as the essence of this problem is about trying to make global extensions of local information. For example, we might naïvely hope to be able to extend a local first integral at the origin to a global first integral. This is not possible in general, but even if we could do so, the first integral would certainly ramify as a global object. Our desire would then be to read off some important information about the system from this global ramification. An example of this is given in Section 10.2.

Over the next four chapters we present several topics very loosely connected with this idea. In this chapter we give two basic and classical examples of monodromy, and then discuss an extended example related to the Model problem of Briskin, Françoise and Yomdin. This is a problem which has close connections with the center-focus problem.

6.1 Some Basic Examples

Our first example is the monodromy of an algebraic function of one variable, $x(c)$ say. The special case where $x(c)$ satisfies $f(x) = c$, for some polynomial f, will be used in the final section and in Chapter 8.

Example 6.1. Suppose, we have a polynomial $F(x, c)$ in $\mathbb{C}[x, c]$. Let $\Delta(c)$ denote the discriminant of F as a polynomial in x over $\mathbb{C}(c)$, and n the degree of F with respect to x. In a neighborhood of each value of c where $\Delta(c) \neq 0$, we can solve $F(x, c) = 0$ to obtain n distinct roots $x_i(c)$, algebraic over c. As these roots do not coalesce unless $\Delta(c) = 0$, we have a locally constant picture as c varies in $\mathbb{C} - S$,

© The Author(s), under exclusive license to Springer Nature Switzerland AG 2024
C. Christopher et al., *Limit Cycles of Differential Equations*, Advanced Courses
in Mathematics - CRM Barcelona, https://doi.org/10.1007/978-3-030-59656-9_6

where S is the set of c where $\Delta(c) = 0$. That is, although the precise value of the x_i will change, of course, there is a unique way of associating the x_i's at each value of c on a path in $\mathbb{C} - S$. However, moving c around a loop which includes a point in S, we find that the x_i will swap amongst themselves in general.

If we fix a base point c_0, and consider the effect on the roots $x_i(c_0)$ as we move around non-trivial loops in $\mathbb{C}-S$ based at c_0, we get a map from $\pi_1(\mathbb{C}-S, c_0)$ to the symmetric group on n elements. This is called the monodromy group of the algebraic function given by $F(x, c) = 0$. We will usually drop the reference to the base point c_0.

It can be shown (for example in [64]) that the monodromy group is equivalent to the Galois group of $\mathbb{C}(x_1(c), \ldots, x_n(c))$ over $\mathbb{C}(c)$.

Our second example examines what happens to the topology of the level curves of a polynomial as a parameter varies. It will be used and generalized in the next two chapters.

Example 6.2. Consider the level curves of the function $x^2 + y^2$. That is, we consider $x^2 + y^2 = c$ as c varies. If we draw this as a two-sheeted covering of $\mathbb{C}$, then we have branch points at $\pm\sqrt{c}$ and can consider a cut lying between these two points. As c makes a loop around $c = 0$ these two branch points swap over.

Now consider two curves. One δ is the loop which surrounds the two branch points, and the other δ' is a loop which begins at ∞ on one sheet, passes through the cut and ends at ∞ on the other sheet. We let (δ, δ') denote the intersection number of δ and δ'.

As c makes the tour around $c = 0$, δ is taken to itself, but δ' tends to $\delta' + (\delta', \delta)\delta$. This is a simple example of the Picard–Lefshetz formula. The transformation on the surface itself (a cylinder) is called a *Dehn twist* (see Figure 6.1).

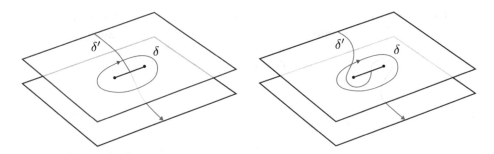

Figure 6.1: A dehn twist.

In the case, for example, when the polynomial is cubic, the generic level curve is a torus, with points added in to represent the intersection of the curve with infinity. The Dehn twist then can be pictured as a non-trivial twist of the torus. This transformation cannot be realised by twisting a torus in three dimensional space, but requires the four-dimensional nature of $\mathbb{C}^2$.

6.2   The Model Problem

We now discuss a longer example where monodromy is used to tackle a problem which is closely related to the center-focus problem.

We consider the Abel equation

$$\frac{dy}{dx} = p(x)y^2 + q(x)y^3, \qquad a \le x \le b \tag{6.1}$$

where p and q are polynomials and a is a fixed constant, We denote the solution of (6.1) by $y(x,c)$, where $y(a,c) = c$. Standard existence theorems ensure that $y(x,c)$ is well defined and analytic in both its arguments for c sufficiently small. If $y(b,c) = c$, then we call $y(x,c)$ a periodic solution. Likewise if $y(b,c) \equiv c$ for all c close to 0 we say that the system has a *center* between a and b. The numbers a and b are not important; by a simple transformation we can always choose $a = 0$ and $b = 1$.

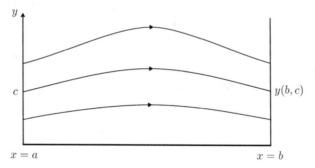

Figure 6.2: The map from $x = a$ to $x = b$.

It is clear that this has close connections with the center-focus problem. We note that one strong motivation for studying Abel equations is that the family of systems,

$$\dot{x} = -y + M(x,y), \qquad \dot{y} = x + N(x,y), \tag{6.2}$$

where M and N are homogeneous polynomials of the same degree n, can be brought to the form (6.1) with p and q trigonometric polynomials. Setting $a = 0$ and $b = 2\pi$, the definitions of periodic solution and center for (6.1) coincide with their usual definitions in the planar system (6.2).

The analogue of the center focus problem in this case is to see what conditions the existence of a center in (6.1) imposes on the defining equations. We shall always denote the antiderivative of the polynomials p and q as P and Q. That is:

$$P(x) = \int_a^x p(\xi)\, d\xi, \qquad Q(x) = \int_a^x q(\xi)\, d\xi.$$

The question of whether there is a center or not depends on computing the expansion of $y(1,c) = \sum \alpha_i c^i$. This can be done as in the case of a center, but involves similar difficulties to the center-focus problem. In the light of this, Briskin, Françoise and Yomdin [21, 22, 23] proposed a simplified center problem called the *model problem*.

Here we ask for conditions that the center lies in a parametric family of centers of the form

$$\frac{dy}{dx} = p(x)y^2 + \epsilon q(x)y^3, \qquad 0 \le x \le 1, \tag{6.3}$$

and has a center for all ϵ sufficiently small. The choice of $a = 0$ and $b = 1$ is done for convenience and is not essential.

A study of the return map calculations described above shows that this condition is equivalent to

$$P(1) = P(0) = 0, \qquad \int_0^1 P(x)^n q(x)\,dx = 0, \qquad n = 0, 1, \dots. \tag{6.4}$$

The problem is then to find out which P and Q can satisfy these equations.

One simple condition, which guarantees that we have a center, is that P and Q are both polynomials of a polynomial A with $A(0) = A(1)$. This is equivalent to the symmetry condition discussed in Chapter 4. We say that the center satisfies the *composition condition* in this case.

In fact, from the form of the return map, this problem turns out to be the same as saying that the center at $\epsilon = 0$ is not destroyed to first order by perturbing ϵ.

We will use some simple ideas using monodromy to show that, if we assume that both $p(0)$ and $p(1)$ are non-zero, then this symmetry condition comprises all cases of centers for the model problem. That is, all centers of this form satisfy the *composition condition*.

6.3   Applying Monodromy to the Model Problem

Let us assume that (6.4) holds with the additional assumption that $p(0)$ and $p(1)$ are non-zero. Without loss of generality we can take P to be monic. Furthermore, if (6.4) holds for some polynomial $P(x)$, then it must also hold for any polynomial of the form $P(x)+k$, with k a constant; whence we can also assume that $P(0) = 0$. The $n = 0$ equation of (6.4) then implies that $Q(0) = Q(1)$. In fact the condition $P(0) = P(1)$ is also deducible from the other conditions.

It is clear that the polynomial $P - c$ can have no roots in the interval $[0, 1]$ for all c such that

$$|c| > K := \sup_{x \in [0,1]} |P(x)|.$$

Thus, if $|c| > K$, $(P(x) - c)^{-1}$ has a well-defined expansion, and

$$I(c) = \int_0^1 \frac{q\,dx}{P(x) - c} = \sum_{i \ge 0} \frac{1}{c^{i+1}} \int_0^1 P(x)^i q(x)\,dx.$$

The hypothesis (6.4) is therefore equivalent to the condition

$$I(c) \equiv 0, \qquad |c| \gg 0.$$

Note that $I(c)$ is related to the Melnikov function which describes the bifurcation of periodic solutions of (6.3) for small ϵ.

We shall work over $\mathbb{C}$ from now on. Let $S = \{c_0, c_1, \ldots, c_n\}$ be the critical values of $P(x)$; that is the values of $P(x)$ when $P'(x) = 0$. In addition we shall also include in S the values $c_0 = P(0) = 0$ and $c_1 = P(1)$, even if they are not critical. For all values of $c \in \mathbb{C} - S$ the polynomial $P(x) - c$ has distinct roots none of which are 0 or 1.

Let $\alpha_i(c)$ be the roots of the polynomial equation $P(x) = c$. Clearly these functions are well defined and non-ramified on the universal cover of $\mathbb{C} - S$, which we denote $\tilde{\mathbb{C}}$, with projection $\pi : \tilde{\mathbb{C}} \to \mathbb{C} - S$. We also lift c and $I(c)$, $(|c| > K)$ to $\tilde{\mathbb{C}}$. We shall show that $I(c)$ has a well-defined expression over $\tilde{\mathbb{C}}$ which, as it vanishes identically in some domain, must vanish throughout $\tilde{\mathbb{C}}$. The proof that all centers satisfy the *composition condition* will follow from an examination of the monodromy of this expression.

Clearly, on $\tilde{\mathbb{C}}$,

$$P(x) - c = \prod (x - \alpha_i).$$

We therefore obtain the following partial fraction expansion over $\tilde{\mathbb{C}}$:

$$\frac{q(x)}{P(x) - c} = r(x, c) + \sum_i \frac{m(\alpha_i(c))}{x - \alpha_i(c)},$$

where r is a polynomial in x and c and $m(x) = q(x)/p(x)$.

Using this expression, we fix a point $c' \in \tilde{\mathbb{C}}$ with $|\pi(c')| > K$ and find that

$$I(c) = R(c) + \sum_i m(\alpha_i) \ln_i(1 - \frac{1}{\alpha_i}), \tag{6.5}$$

in a neighborhood of c'. Here R is a polynomial in c, and the $\ln_i$'s are specific branches of the logarithm, chosen individually for each α_i. It is clear that each term of this expression can be analytically continued to $\tilde{\mathbb{C}}$ as the only points of indeterminacy are those where the algebraic functions $\alpha_i(c)$ are ramified or where one of the $\alpha_i(c)$ attains the value 0 or 1. Both of these possibilities have been taken care of by removing the set S from consideration.

It is an interesting fact that the remainder of the proof does not depend on the specific choice of these logarithmic branches.

From standard results on algebraic functions, we know that for c sufficiently close to 0, we can describe all the solutions of $P(x) - c$ as Puisseaux expansions in c. We label the distinct roots (except $x = 1$) of $P(x) = 0$ as x_i, $i = 0, \ldots, r - 1$. Since $P(0) = 0$, we can choose $x_0 = 0$ without loss of generality. If we also have $P(1) = 0$, then we shall label this root x_r. The contribution of this term in the expressions derived below should be taken to be zero if $P(1) \neq 0$. We write l_i for the multiplicity of x_i as a root of $P(x) = 0$.

About the point $x = x_i$ we have

$$c = k_i(x - x_i)^{l_i}(1 + U_i(x - x_i)), \qquad U_i(0) = 0,$$

for some constant $k_i \neq 0$ and some polynomial U_i. Close to $c = 0$, therefore, the solutions of $P(x) = c$ can be written as

$$x_{i,j} - x_i = \zeta_i^j k_i^{-1/l_i} c^{1/l_i}\left(1 + h_i(\zeta_i^j c^{1/l_i})\right), \quad i = 0, \dots, r, \quad j = 0, \dots, l_i - 1$$

where h_i is analytic with $h_i(0) = 0$, and ζ_i is a primitive l_i-th root of unity. We fix the choice of k_i^{-1/l_i} for each i. Clearly $x_{i,j}(0) = x_j$.

If we now analytically continue the expression (6.5) to a neighborhood of $c = 0$, then the solutions α_i of $P(x) = c$ can be relabeled as the corresponding $x_{ij}(c)$. In the same way we write $\ln_{i,j}$ for the corresponding branch of the logarithm given in (6.5). Thus,

$$I(c) = R(c) + \sum_{i=0}^{r} \sum_{j=0}^{l_i-1} m(x_{i,j}) \ln_{i,j}(1 - \frac{1}{x_{i,j}}). \tag{6.6}$$

For ease of explanation, we shall evaluate the second index of $x_{i,j}$ modulo l_i. Clearly, as $c \in \mathbb{C} - S$ moves around a sufficiently small circle about 0, γ say, the $x_{i,j}$, $j = 0, \dots, l_i - 1$ move around x_i swapping roles as follows:

$$x_{i,j} \mapsto x_{i,j+1}, \qquad (j = 0, \dots l_i - 1).$$

Alternatively, we can consider the effect of the corresponding deck transformation of $\tilde{\mathbb{C}}$ on the right-hand side of (6.6). We denote this transformation by σ. Thus

$$\sigma^k \ln_{i,j}(1 - \frac{1}{x_{i,j}}) = \ln_{i,j}(1 - \frac{1}{x_{i,j+k}}).$$

If γ is sufficiently small, the paths of the $x_{i,j}$, $i \neq 0, r$ will not encircle the values 0 or 1. Thus, applying σ^{l_i} to $\ln_{i,j}(1 - 1/x_{i,j})$ will return it to its original value.

However, when $i = 0, r$, the roots $x_{0,j}$ cycle around 0 and 1 and so

$$\sigma^{l_0} \ln_{0,j}(1 - \frac{1}{x_{0,j}}) = \ln_{0,j}(1 - \frac{1}{x_{0,j}}) - 2\pi i,$$

$$\sigma^{l_r} \ln_{r,j}(1 - \frac{1}{x_{r,j}}) = \ln_{r,j}(1 - \frac{1}{x_{r,j}}) + 2\pi i.$$

Under the assumption that $p(0)$ and $p(1)$ are non-zero, we must have $l_0 = 1$ and $l_r \leq 1$. Letting N be the lowest common multiple of the l_i we obtain

$$\sigma^N I(c) - I(c) = -2\pi i\, N\, m(x_{0,0}) + 2\pi i\, N\, m(x_{r,0}).$$

Now $\sigma^N I(c) \equiv 0 \equiv I(c)$ and so $m(x_{0,0}) = m(x_{r,0})$. Since $q(x_{0,0})$ cannot vanish, the term $x_{r,0}$ must also exist and so we deduce $l_0 = 1$ and $P(1) = 0$.

Finally, we consider

$$M(c) = Q(x_{0,0}) - Q(x_{r,0}).$$

It is easy to see that

$$M'(c) = m(x_{0,0}) - m(x_{r,0}) = 0, \qquad M(0) = 0,$$

and so $M(c)$ is identically zero. We therefore have two polynomials P and Q for which their values on $x_{0,0}(c)$ and $x_{r,0}(c)$ are the same. The set of all rational functions for which this is true forms a subfield $\mathcal{F}$ of $\mathbb{C}(x)$, Following an identical line of reasoning to that used in Chapter 4 we see that both P and Q must be polynomials of a polynomial A in $\mathcal{F}$. Furthermore, since $A(x_{0,0}) = A(x_{r,0})$, we must have $A(0) = A(1)$. This proves that the composition condition is a necessary and sufficient condition for a center in this case $(p(0), p(1) \neq 0)$.

Notes

§1 The results on the model problem presented here first appeared in [38]. It was originally conjectured that the *composition condition* would hold for all systems (6.1) with centers. However, Pakovich [100] gave examples which showed that this was not the case. Although there is no global composition, the example is given by a sum of terms which are given by composition. In this case, we need to find a P which can be decomposed in more than one way.

This can be done, for example, by the use of Chebyshev polynomials. These are polynomials, $T_n(x)$, defined such that

$$T_n(\cos(x)) = \cos(nx).$$

(see also Definition 8.9). These satisfy the following commutation property:

$$T_m \circ T_n = T_{mn} = T_n \circ T_m.$$

For ease, we consider the model problem with $a = -\sqrt{3}/2$ and $b = \sqrt{3}/2$. If we take, for example,

$$P(x) = T_6(x), \quad \text{and} \quad Q = T_3(x) + T_2(x),$$

then $P(a) = P(b)$ and $Q(a) = Q(b)$ and

$$\int_a^b P(x)^n \, q(x) \, dx = \int_a^b T_6(x)^n \, T_2(x) \, dx + \int_a^b T_6(x)^n \, T_3(x) \, dx.$$

Both of the terms on the right hand side are zero for all non-zero n, as they are compositions of T_2 and T_3 respectively, but P and Q do not have a common composition.

The decomposition property of Chebyshev polynomials is rather special. In fact Ritt's theorem [107] characterises all polynomials which can be decomposed in more than one way as essentially (modulo compositions with linear maps) powers, Chebyshev polynomials, and terms of the form:

$$x^{rm} f(x^m)^m = x^m \circ (x^r f(x^m)) = (x^r f(x)^m) \circ x^m.$$

§2 Much work has been carried out on Abel equations following the works [21, 22, 23]. In particular, consideration of the effect of extending the local return map to obtain global conditions (see, for example [24, 66]). We shall return to this idea in Section 10.2 where we give a simple criterion, using ideas from Chapter 4, under which the center of (6.1) must be due to composition.

§3 In [102], Pakovich and Musychuk gave a complete solution of the polynomial moment problem in terms of sums of compositions as above. In particular, they show that Q can be decomposed as a sum

$$Q(x) = \sum_{i=1}^{r} Q_i(x), \quad \text{where} \quad Q_i(x) = \tilde{Q}_i(W_i(x)), \quad W_i(a) = W_i(b), \qquad (6.7)$$

and where, for each i, we can write $P(x) = \tilde{P}_i(W_i(x))$. The proof uses elements of representation theory, in particular the theory of Schur rings (see, for example, [118, 126]).

Pakovich, in [101], applied this result to settle the composition conjecture for parametric centers. In this case, the behaviour for large values of the parameter give a complementary set of moment conditions to equations (6.4):

$$\int_0^1 Q(x)^n p(x)\, dx) = 0, \qquad n \geq 0.$$

Applying the results for [102] to (6.4) and the equation above, shows that the composition conjecture is satisfied in this case.

§4 Since the polynomial case is a model for the trigonometric situation, it is interesting to see how much of these ideas can be applied in this case to obtain information about the center-focus problem for the homogenous systems (6.2).

Writing the trigonometric polynomials as functions of $z = e^{i\theta}$, leads to considering the following variation of the moment problem: describe the Laurent polynomials $P(z)$ and $Q(z)$ such that

$$\oint_{|z|=1} P(z)^n\, dQ(z) = 0, \qquad n \geq 0. \tag{6.8}$$

In [1], the trigonometric moment problem is studied in some detail. Clearly, the composition of Laurent polynomials, corresponding to (6.7) will satisfy (6.8), but we also have another case which is sufficient to give (6.8), namely:

$$P(z) = \tilde{P}(z^\ell), \qquad Q(z) = \tilde{Q}(z^\ell) + \sum_{\ell \nmid i} a_i z^i, \tag{6.9}$$

where $\tilde{P}$ and $\tilde{Q}$ satisfy

$$\oint_{|z|=1} \tilde{P}(z)^n \, d\tilde{Q} = 0, \qquad n \geq 0. \tag{6.10}$$

In order to prove (6.10), we can again apply one of the two mechanisms above. Since $\tilde{P}$ and $\tilde{Q}$ are of lower degree, this process must terminate at some point.

By considering the Galois group over $\mathbb{C}(c)$ generated by the equation $P(z) = c$, the moment problem can be recast in terms of the action of the Galois group on the roots. Using GAP, it was possible to show that, except for a small number of exceptional groups, that for all Galois groups up to order 30, the solutions of (6.8) are obtained by iterating combinations of the two mechanisms above.

Restricting to real trigonometric polynomials, it is possible to show that the exceptional groups do not occur, and therefore for all Laurent polynomials of degree 15 or less, all solutions of the trigonometric moment problem:

$$\int_0^{2\pi} P(\cos(\theta), \sin(\theta))^n \, dQ(\cos(\theta), \sin(\theta)) = 0, \qquad n \geq 0,$$

must be iterations of the following two mechanisms:

1. There exist trigonometric polynomial $\tilde{P}$ and $\tilde{Q}$ and an integer $\ell > 1$, such that

   $$P = \tilde{P}(\cos(\ell\theta), \sin(\ell\theta)), \qquad Q = \tilde{Q}((\cos(\ell\theta), \sin(\ell\theta))) + R(\cos(\theta), \sin(\theta)),$$

   where the fourier expansion of R contains no terms of the form $\cos(k\theta)$ or $\sin(k\theta)$ for k a multiple of ℓ, and where $\tilde{P}$ and $\tilde{Q}$ also satisfy

   $$\int_0^{2\pi} \tilde{P}(\cos(\theta), \sin(\theta))^n \, d\tilde{Q}(\cos(\theta), \sin(\theta)) = 0, \qquad n \geq 0.$$

2. There exist polynomials P_i and Q_i and trigonometric polynomials W_i such that

   $$P = P_i(W_i(\cos(\theta), \sin(\theta))), \qquad Q = \sum_{i=1}^r Q_i(W_i(\cos(\theta), \sin(\theta)).$$

It is conjectured that iterations of these two mechanisms comprise all real solutions to the trigonometric moment problem.

§5 It was an open question for a while whether a polynomial Abel equation could have centers given by Darboux type integrals. Such systems have now been found [71]. The example has p and q of degree 4 and 9 respectively, and a first integral of the form $H = x^2 f_1^3 / f_2^4$, where f_1 and f_2 are polynomials in x and y of degree 12. It is not clear whether the Darboux centers are just exceptional, or it is just that their existence requires a high degree of system.

Problem 6.1. Investigate Darboux centers for Abel equations. Do they become more prevalent as the degree of the system increases? Are there other mechanisms which underly the existence of a center for Abel equations?

§6 Ideas of monodromy are ubiquitous in mathematics occurring even in places where they would seem unlikely (for example they play a key role in arithmetic algebraic geometry).

Żołądek's book [135] contains much more detail on many topics associated to monodromy in differential equations, and further afield. The book by Il'Yashenko and Yakovenko [77] is also very helpful.

Chapter 7

The Tangential Center-Focus Problem

As is well known, the second part of Hilbert's 16th problem is concerned with bounding the number of limit cycles in a polynomial system (1.2) of degree n in terms of n. This is a very hard problem, but Arnold has suggested a "Weak Hilbert's 16th problem" which seems far more tractable: to find a bound on the number of limit cycles which can bifurcate from a first-order perturbation of a Hamiltonian system,

$$\dot{x} = -\frac{\partial H}{\partial y} + \epsilon P, \qquad \dot{y} = \frac{\partial H}{\partial x} + \epsilon Q, \tag{7.1}$$

where the Hamiltonian, H, is a polynomial of degree $n + 1$ and the perturbation terms, P and Q are polynomials of degree m.

Over time, the term "Tangential Hilbert's 16th problem" seems to have become the more popular (and descriptive) phrase for this problem. Although this is still a very difficult problem, finiteness results are known in this case, and several exact results have been obtained for Hamiltonians of low degree.

Recall that, to first order, the limit cycles appearing in the perturbed system (7.1) are given by the zeros of the abelian integral

$$I_c = \oint_{\gamma_c} P \, dy - Q \, dx, \tag{7.2}$$

where γ_c is a family of closed loops in the level curves, $H = c$, of the Hamiltonian.

We want to suggest here an analogous "tangential center-focus problem", which seems to have a similar property of being much easier to tackle, whilst retaining something of the structure of the original problem.

Suppose that the Hamiltonian H has a Morse point at the origin. That is, $\partial H/\partial x = \partial H/\partial y = 0$ at $(0,0)$ and the matrix of second derivative of H is sign-definite. Then (taking $-H$ if necessary) we can write $H = H(0,0) + X^2 + Y^2$ for some suitable choice of local coordinates, and the system (7.1) clearly has a center at the origin for $\epsilon = 0$. Without loss of generality we can take $H(0,0) = 0$. The

C. Christopher et al., *Limit Cycles of Differential Equations*, Advanced Courses in Mathematics - CRM Barcelona, https://doi.org/10.1007/978-3-030-59656-9_7

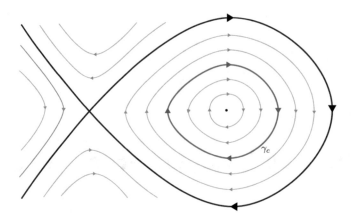

Figure 7.1: A level curve γ_c in a Hamiltonian system.

curves $X^2 + Y^2 = c$, for c close to zero, give a family of closed curves tending to the origin as c tends to zero. We call such curves *vanishing cycles*.

We call the origin a *tangential center* if

$$\oint_{\gamma_c} P\,dy - Q\,dx \equiv 0, \tag{7.3}$$

for γ_c the vanishing cycle $X^2 + Y^2 = c$ given above, and c sufficiently small.

The *tangential center-focus problem* then asks for the conditions on P and Q which give a tangential center.

We shall sketch a proof in this chapter that for a generic Hamiltonian the answer for this problem is quite simple: P and Q must satisfy the equation

$$P\,dy - Q\,dx = dA + B\,dH, \tag{7.4}$$

for some polynomials A and B in x and y. In this case we say that the 1-form $P\,dy - Q\,dx$ is *relatively exact*.

The proof is due to Il'yashenko [76] who used the result to count the dimension of the space of non-trivial perturbations of the Hamiltonian system (7.1) and hence give a lower bound on the maximum number of limit cycles which can be produced by such perturbations.

Françoise [65] has shown that if the Hamiltonian has the property that every tangential center must come from a relatively exact perturbation, then it is possible to calculate the higher order perturbation terms of (7.1) as abelian integrals. This is known not to be true in general. Françoise calls this property "condition (∗)".

The interesting question is how degenerate a Hamiltonian needs to become before condition (∗) does not hold. We shall show in the next chapter that, in the case of hyperelliptic Hamiltonians, the failure of condition (∗) implies the existence of a non-trivial symmetry.

7.1 Preliminaries

Suppose we have a tangential center (7.3) and want to investigate its global consequences. Since (7.2) is analytic in c, we can extend the condition (7.3) up to the boundary of the period annulus around the origin. However, if we want to extend further, we can only do this by working over the complex numbers.

If we do so, the level curves $H(x, y) = c$ become Riemann surfaces, and the curves γ_c closed curves on this surface. The equations (7.2) and (7.3) carry across in an obvious way. By Cauchy's theorem we only need to consider the curves γ_c up to homotopy.

If we exclude a number of *critical values*, $S = \{c_1, \ldots, c_r\}$, the map $(x, y) \mapsto H(x, y)$ defines a fibration over $\mathbb{C} - S$, with fibre $H(x, y) = c$ of constant topological type. If we consider the loops γ_c only up to homotopy, then there is a unique way to transport γ_c as c varies.

If we move around a non-trivial loop in $\mathbb{C} - S$, the curve γ_c will not in general return to itself, but some other element in the homology group of the Riemann surface.

Thus we get a map

$$\sigma : \pi_1(\mathbb{C} - S) \to Aut(H_1(\phi_c, \mathbb{Z})), \tag{7.5}$$

where $H_1(\phi_c, \mathbb{Z})$ denotes the first homology group of the Riemann surface ϕ_c, lying in the level curve $H(x, y) = c$ with $\mathbb{Z}$ coefficients.

Our aim will be to show that for a generic Hamiltonian we can generate the whole homology group $H_1(\phi_c, \mathbb{Z})$ from the one cycle γ_c. By this we mean the following.

Definition 7.1. We say that γ_c generates the homology $H_1(\phi_c, \mathbb{Z})$ if there exist loops ℓ_i, $i = 1, \ldots, l_k$, in $\pi_1(\mathbb{C} - S)$ such that

$$\sum_{i=1}^{k} \mathbb{Z} \sigma(\ell_i) \gamma_c = H_1(\phi_c, \mathbb{Z}).$$

Similarly, we say that γ_c generates $H_1(\phi_c, \mathbb{Q})$ if we can find ℓ_i as above such that

$$\sum_{i=1}^{k} \mathbb{Q} \sigma(\ell_i) \gamma_c = H_1(\phi_c, \mathbb{Q}).$$

We first consider a simple case. Take the Hamiltonian

$$H = y^2 + x^2 + x^3.$$

This has critical points at $(0, 0)$ and $(-2/3, 0)$. The critical values where the topology of $H = c$ changes are therefore $c_1 = 0$ and $c_2 = 4/27$.

Take γ_c as the vanishing cycle at the origin for $0 < c \ll 1$ close to 0. We now move c along a path which makes a positive loop around $c = c_2$ and then $c = c_1$.

At both the critical values there is a Dehn twist. Let γ_c' be the vanishing cycle at $c = c_2$, then the twist at $c = c_2$ takes γ_c to $\gamma_c + \gamma_c'$. Then the twist at $c = c_1$ takes $\gamma_c + \gamma_c'$ to γ_c'. Since γ_c and γ_c' form a basis for $H_1(\phi_c, \mathbb{Z})$, we have shown that γ_c generates $H_1(\phi_c, \mathbb{Z})$.

7.2 Generic Hamiltonians

We want to generalize the construction in the previous section. We follow the original paper of Il'yashenko [76].

Consider the space B of Hamiltonians of degree $n + 1$ with only Morse singularities, with distinct critical values, and whose highest order terms have no multiple factors. Equivalently, we require the equations $\partial H / \partial x = 0$ and $\partial H / \partial y = 0$ to have n^2 distinct solutions and the values of H on these solutions to also be distinct. These conditions are generic, and hence the complement of B in the space of all polynomials of degree $n + 1$, $\mathbb{C}^{(n+2)(n+3)/2}$ is a proper algebraic subset. In particular B is path connected.

Now it is clear by continuity that if we are in the space B, the property that a vanishing cycle γ_c generates the homology is preserved under perturbation, so that if it holds for any Hamiltonian in B it holds in all of them.

Let $S = \{c_1, \ldots, c_k\}$ be the critical values as above, and denote by $\tilde{C}$ the universal cover of $C = \mathbb{C} - S$.

Near to each critical value c_i we have a vanishing cycle δ_i (unique up to homotopy). We fix a point $\tilde{c}_0 \in \tilde{C}$ with image $c_0 \in C$ and paths l_i in C from a sufficiently small neighborhood of each of the c_i to c_0 in C. We denote the result of transporting δ_i along l_i by $\delta_i(\tilde{c}_0)$ and can extend this to a cycle $\delta_i(\tilde{c})$ for all $\tilde{c} \in \tilde{C}$.

We say that the set of vanishing cycles is good if the cycles $\delta_i = \delta_i(\tilde{c}_0)$ generate the homology $H_1(\phi_{c_0}, \mathbb{Z})$ and we can connect any two cycles, d_i and d_j by a chain of cycles

$$\delta_i = \delta_{n_1}, \delta_{n_2}, \ldots, \delta_{n_k} = \delta_j,$$

with $(\delta_{n_r}, \delta_{n_{r+1}}) = \pm 1$ for each r. Here, (δ', δ) denotes the intersection number of the curves δ and δ' on the Riemann surface ϕ_c.

Clearly, if a Hamiltonian $H \in B$ has a good set of cycles for $\tilde{c} = \tilde{c}_0$, they will also be good for every value of $\tilde{c}$. Furthermore they will transport in an obvious way to a good set of cycles for any Hamiltonian in B.

If we have a good set of cycles, then it is now easy to show that γ_c generates the homology $H_1(\phi_{c_0}, \mathbb{Z})$. Suppose we have two vanishing cycles δ_i and δ_j with $(\delta_i, \delta_j) = \pm 1$. For each i, let $\bar{l}_i$ be the path from c_0 to itself which moves along l_i^{-1}, around c_i in the positive sense and back along l_i to c_0. Then $\sigma(\bar{l}_j)$ takes δ_i to

$$\delta_i + (\delta_j, \delta_i)\delta_j,$$

and $\sigma(\bar{l}_i)$ takes this to

$$\delta_i + (\delta_j, \delta_i)\delta_j + (\delta_i, \delta_i + (\delta_j, \delta_i)\delta_j)\delta_i = \pm\delta_j,$$

where equality is taken in $H_1(\phi_{c_0}, \mathbb{Z})$. Thus, since any two vanishing cycles can be connected by a chain of cycles with intersection numbers ± 1, we can generate the homology $H_1(\phi_{c_0}, \mathbb{Z})$ from any vanishing cycle.

To finish, we sketch the proof that we can find a good set of cycles. The idea is to consider a Hamiltonian

$$\tilde{H} = \prod_{i=1}^{n+1} r_i,$$

given by the product of $n + 1$ real linear factors, such that the lines $r_i = 0$ are in general position. This Hamiltonian does not lie in the space B as the critical value $c = 0$ is very degenerate. However, we do not need that all the critical values are distinct to define a good system of cycles. Since this system lies in the complement of B it can be moved into B by a perturbation which will maintain the good system of cycles.

Consider the Riemann surface for a general value of c. This is non-singular in the finite plane with $n+1$ points of intersection with infinity and genus $n(n-1)/2$. Thus it has Betti number

$$2\frac{n(n-1)}{2} + (n+1) - 1 = n^2.$$

Now we consider the level curves of $\tilde{H} = c$. For $c = 0$ we have $n(n+1)/2$ points corresponding to the intersections of the $r_i = 0$. The other $n(n-1)/2$ vanishing cycles come from the centers which lie in the bounded connected components of $H > 0$ and $H < 0$. This gives us a full set of generators for the homology (see Figure 7.2).

All critical values lie on the real line. We take c_0 to be a point sufficiently close to the origin and take paths from neighborhoods of the other critical points to c_0 passing $c = 0$ by making a half turn in the positive direction if necessary. It can be readily verified that this will give us a good system of cycles.

Note that even if we have a good system of cycles for $\tilde{H}$, it is not possible to conclude that the vanishing cycles generate the homology, as we require that the critical values be distinct in order to apply the argument above. That is, the result only holds true for the perturbed Hamiltonian.

7.3 Relative exactness

We have shown that we can generate the homology $H_1(\phi_c, \mathbb{Z})$ from the vanishing cycle γ_c. Thus we are left to show that the condition (7.3) implies that the 1-form $P\,dy - Q\,dx$ is relatively exact. However, we have shown that γ_c generates

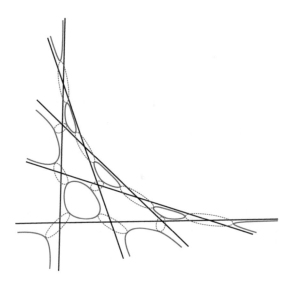

Figure 7.2: A product of lines with perturbation and vanishing cycles (dashed).

$H_1(\phi_c, \mathbb{Z})$, and since (7.2) is analytic in c we must have

$$\oint_\gamma P \, dy - Q \, dx = 0, \tag{7.6}$$

for any closed curve γ in ϕ_c. We say that $P \, dy - Q \, dx$ is *topologically exact* on ϕ_c.

We want to show that if a polynomial 1-form is topologically exact on ϕ_c for all c in a continuum, then the 1-form must be relatively exact. We follow the paper of Bonnet [18].

Proposition 7.2. *If ϕ_c is a smooth irreducible curve, then any 1-form ω which is topologically exact on ϕ_c can be written in the form*

$$\omega = dR + (H - c) \, \eta,$$

for some polynomial R and some polynomial 1-form η.

Proof. The condition (7.6) means that we can integrate the 1-form ω on ϕ_c and get a single-valued function $\bar{R}$. The function R is analytic on ϕ_c, and by considering the growth of R at infinity, it must be a rational function and therefore realizable as the restriction of a rational function f/g in $\mathbb{C}^2$ whose denominator does not vanish on ϕ_c. This latter condition implies that $A(H - c) + Bg = 1$ for some polynomials A and B by the Nullstellensatz, and so the $f/g = Bf$ on ϕ_c. We take $R = Bf$ whose restriction to ϕ is $\bar{R}$. Then $\omega - dR$ vanishes on ϕ_c and hence the result. $\qquad\square$

Proposition 7.3. *Suppose that there exist polynomials R_c and polynomial 1-forms η_c such that*

$$\omega = dR_c + (H - c)\,\eta_c, \tag{7.7}$$

for an uncountable number of parameter values of c; then there exists a polynomial $P(H)$ such that

$$P(H)\,\omega = dA + B\,dH,$$

for some polynomials A and B.

Proof. Consider the space of 1-forms η_c from (7.7). Since the space of polynomial 1-forms is countable, but the number of η_c are uncountable, then there must exist λ_i and c_i, $i = 1, \ldots, r$, such that

$$\sum_{i=1}^{r} \lambda_i \eta_{c_i} = 0.$$

Hence,

$$\sum_{i=1}^{r} \frac{\lambda_i}{H - c_i}\,\omega = \sum_{i=1}^{r} \frac{\lambda_i\,dR_{c_i}}{H - c_i}.$$

On multiplying by the product of the $H - c_i$ and integrating by parts we obtain the desired result. $\qquad\square$

Theorem 7.4. *Suppose that for all c, the curve $\phi_c = \{H = c\}$ is connected and contains only finitely many singular points of H. If a 1-form is relatively exact on ϕ_c for an uncountable number of values of c, then the 1-form is relatively exact.*

Proof. From Proposition 7.3 we only need to show that if $(H - c)\,\omega$ is relatively exact, then so is ω. Thus, suppose that

$$(H - c)\,\omega = dA + B\,dH,$$

for some polynomials A and B. On $H = c$ we have $dA = 0$ thus A is a constant on $H = c$ since $H = c$ is connected. Thus $A = k + (H - c)\bar{A}$ for some constant k and some polynomial $\bar{A}$, and we can write

$$(H - c)\,\omega = (H - c)\,d\bar{A} + (\bar{A} + B)\,dH.$$

Finally, since there are only a finite number of points on $H = c$ where dH vanishes, $(H - c)$ must divide $\bar{A} + B$ and we are done. $\qquad\square$

The hypothesis for this theorem is satisfied if we assume that the Hamiltonian H lies in B and therefore we have shown that for a generic Hamiltonian all tangential centers arise from relatively exact perturbation terms.

Notes

§1 In the paper of Bonnet [18], a more detailed result is shown. The polynomial, $P(H)$, in Proposition 7.3 can be reduced to

$$P(H) = \prod_{i=1}^{r} (H - c_i)^{m_i},$$

where $m_i = 1$ if $H = c_i$ contains infinitely many singular points of H; m_i is arbitrary if $H = c_i$ is not connected; and $m_i = 0$ otherwise.

That more complex things can happen for more general Hamiltonians is clear in the case of the Hamiltonian $\tilde{H}$ mentioned in Section 7.2. In fact, taking

$$P \, dy - Q \, dx = \sum \alpha_i \, K_i \, dr_i, \qquad K_i = \tilde{H}/r_i,$$

for any constants α_i, it is clear that on $\tilde{H} = c$ we have

$$P \, dy - Q \, dx = \sum \alpha_i \, c \, dr_i/r_i,$$

and hence (7.3) holds for each of the real centers lying in the bounded regions of $H > 0$ and $H < 0$. However the form cannot be relatively exact as the integrals around the vanishing cycles for $c = 0$ will not vanish in general. Perturbations from such Hamiltonians have been studied by Movasati [97] and Uribe [122] and, more recently, López García [88].

Such 1-forms, which generalize the relatively exact 1-forms, seem to be the tangential analog of the Darboux centers. We shall show in the next section that the symmetric centers also have a prominent place in the tangential center-focus problem. Thus although the tangential center-focus problem is a considerable simplification of the original center-focus problem, much of the complexity of the full center-focus problem is retained.

§2 In general, if the first order perturbation is identically zero, we then need to consider the perturbations to higher order. In the case when the Hamiltonian is sufficiently generic, Françoise has shown how Condition (∗) can be used to simplify these higher order computations [65].

In particular, if the first k terms of the Hamiltonian perturbation are identically zero, then the next term can be expressed as an Abelian integral.

§3 It is possible to generalize the essence of Il'yashenko's result to systems with generic Darboux centers [55]. Suppose the system has a Darboux center with first integral (written in logarithmic form)

$$F = \sum_{i=1}^{r} \lambda_i \ln(f_i(x, y)).$$

If M represents the product of the curves f_i, then the perturbed system can be written as

$$\dot{x} = -MF_y + \epsilon A, \qquad \dot{y} = MF_x + \epsilon B. \tag{7.8}$$

The level curves of F once again give a family of vanishing cycles γ_c close to the center. However, in order to apply the theory of the previous sections, we must divide the system by M to get a Hamiltonian system before perturbation.

The bifurcation of limit cycles from (7.8) is therefore given by a *pseudo-Abelian* integral,

$$I_c = \oint_{\gamma_c} \frac{A\,dy - B\,dx}{M}.$$

We wish to know what we can deduce about the perturbation terms, A and B if I_c is identically zero close to the center.

We say that $A\,dy - B\,dx$ is *Darboux relatively exact* if it satisfies the equation

$$\frac{A\,dy - B\,dx}{M} = \frac{P}{M}\frac{dF}{F} + d\left(\frac{R}{M}\right) + \sum_{i=1}^{r} \alpha_i \frac{df_i}{f_i},$$

for some polynomials P and R and constants α_i. If $A\,dy - B\,dx$ are relatively exact, and the curves γ_c do not encircle the curves $f_i = 0$ (which is the case when the centre does not lie on any of the curves), then it is clear that the integral I_c must vanish identically.

In order to prove that, conversely, the vanishing of I_c locally implies that the perturbation must be Darboux relatively exact, we must make some generic assumptions on the Darboux integral.

First, we assume that the union of the curves $f_i = 0$ together with the line at infinity must form a normal crossing hat $M = 0$. Secondly, we assume that the critical values of F, except 0, are distinct. Lastly, we assume that each ratio of eigenvalues, λ_i/λ_j should be irrational as well as the ratios λ_i/Λ, where $\Lambda = \sum \lambda_i$. The latter ratios are included to take account of the ratio of eigenvalues with the line at infinity.

As c varies, the curves γ_c undergoes a more complex variation than in the Hamiltonian case, due to the multivalued nature of the Darboux integral. However, in the neighbourhood of each of the curves $f_i = 0$ there are well-defined cycles. These are the Pochhammer cycles obtained by winding around the intersection of $f_i = 0$ with any other two curves.

Under the generic assumptions above, it turns out that the vanishing of the I_c locally can be shown to imply the vanishing of the integral around any pochhammer cycle in the neighbourhood of a curve $f_i = 0$. This, in turn, allows one to construct a global integral in the neigbourhood of a curve which can then be extended to $\mathbb{C}^2$.

The full proof of the statements above is quite involved, but has a number of strong consequences. First, it establishes, in a direct manner, Movasati's result that families of generic Darboux centers form complete components of the center variety.

Secondly, by some simple counting arguments, it is possible to calculate the number of limit cycles bifurcating from generic Darboux centers. In particular, a Darboux centre given by $d + 1$ invariant lines can bifurcate $d^2 - 2$ limit cycles.

Under the generic assumptions above, a generalisation of Condition $(*)$ also exists and allows one to apply Françoise algorithm to perturbations of Darboux systems to calculate higher order perturbation terms under the assumption that the lower order terms vanish as pseudo-Abelian integrals.

However, in this case, the power of M in the denominator of the pseudo-Abelian integrals obtained increase with the order of ϵ. In order to proceed, the results on Darboux relative exactness above can be extended to include such integrands. In this case, we say that the integrand is Darboux relatively exact if it satisfies

$$\frac{\tilde{A}\,dy - \tilde{B}\,dx}{M^k} = \frac{\tilde{P}}{M^k}\frac{dF}{F} + d\left(\frac{\tilde{R}}{M^k}\right) + \sum_{i=1}^{r}\tilde{\alpha}_i\frac{df_i}{f_i},$$

for some polynomials $\tilde{P}$ and $\tilde{R}$, and constants $\tilde{\alpha}_i$.

Chapter 8

Monodromy of Hyperelliptic Abelian Integrals

We want to show that in the case of Hamiltonians of the form

$$H(x,y) = y^2 + f(x),$$

where $f(x)$ is a polynomial of degree n, the existence of a tangential center implies that either $P\,dx - Q\,dy$ is relatively exact, or the polynomial $f(x)$ is *composite*. That is, it can be expressed as a polynomial of a polynomial, $f(x) = a(b(x))$, in a non-trivial way.

The factorization is related to the existence of a symmetry in the Hamiltonian. That is, a factorization

$$H(x,y) = \tilde{H}(r(x,y), s(x,y)), \qquad (8.1)$$

where $r(x,y)$ and $s(x,y)$ are polynomials whose Jacobian vanishes at some point in $\mathbb{C}^2$, and hence on some algebraic curve C.

Conversely, suppose we are given a factorization of a Hamiltonian H as in (8.1). Let $\pi : \mathbb{C}^2 \to \mathbb{C}^2$ be the projection $(x,y) \mapsto (r(x,y), s(x,y))$. Suppose further that we have a family of vanishing cycles γ_c whose image under π is homotopic to zero. Let $\tilde{w}$ be a polynomial 1-form on $\mathbb{C}^2$, and $w = \pi^*\tilde{w}$, then

$$\oint_{\gamma_c} w = \oint_{\pi(\gamma_c)} \tilde{w} \equiv 0,$$

thus the 1-form w gives a tangential center. However, in general, this 1-form is not relatively exact, since any family of vanishing cycles which are not sent to a curve homotopic to zero can have non-trivial integrals given by the integral of $\tilde{w}$.

In this section we want to show the following result.

© The Author(s), under exclusive license to Springer Nature Switzerland AG 2024
C. Christopher et al., *Limit Cycles of Differential Equations*, Advanced Courses
in Mathematics - CRM Barcelona, https://doi.org/10.1007/978-3-030-59656-9_8

Theorem 8.1. *If the system* (7.1), *with* $H = y^2 + f(x)$, *has a tangential center with associated vanishing cycle* γ_c, *then one of the following must hold.*

(i) γ_c *generates the homology* $H_1(\phi_c, \mathbb{Q})$.

(ii) f *is decomposible.*

(iii) f *is a Chebyshev polynomial of prime degree.*

Remark 8.2. Note that the theorem doesn't say anything about the perturbation terms, it is purely topological. However, in generic cases of (i) we can conclude that the perturbation terms would have to be relatively exact. In Case (ii) we show later that the 1-form given by the perturbation terms will be the sum of a relatively exact term and the pull back of a 1-form on the factorized space. In Case (iii) it is possible by analyzing the monodromy in more detail to show that in fact we can reduce to Case (ii). However, in this case we do have a non-trivial splitting of $H_1(\phi_c, \mathbb{Q})$ into invariant subspaces over an extension of $\mathbb{Q}$. It is not clear whether splittings can occur for any other Hamiltonians not in Case (ii).

8.1   Some Group Theory

We recall some definitions from group theory.

Definition 8.3.   (1) Let G be a group acting on a finite set S; then we say that the action is *imprimitive* if there exists a non-trivial decomposition of S, $S = \bigcup S_i$, such that for each element of g and each i, g sends S_i into S_j for some j. The action is called *primitive* if it is not imprimitive.

(2) The action is *transitive* if, given any pair of elements of S, s_1 and s_2, there is an element $g \in G$ which sends s_1 to s_2.

(3) The action is *2-transitive* if, given any two pairs of elements of S, (s_1, s_2) and (s_3, s_4), there is an element $g \in G$ which sends s_1 to s_3 and s_2 to s_4.

(4) The action is *regular* if, given two elements s_1 and s_2 of S, there is a unique element g of G which sends s_1 to s_2.

Clearly, if G is transitive and imprimitive, then all the sets S_i must be of the same size.

Recall that the affine group $\mathrm{Aff}(\mathbb{Z}_p)$ is the group of all affine transformations of $\mathbb{Z}_p$ to itself. That is, it is the group of all maps from $\mathbb{Z}_p$ to itself of the form $x \mapsto ax + b$ for $a, b \in \mathbb{Z}_p$ with multiplication given by composition.

Theorem 8.4 (Burnside–Schur). *Every primitive finite permutation group containing a regular cyclic subgroup is either 2-transitive or permutationally isomorphic to a subgroup of the affine group* $\mathrm{Aff}(p)$, *where* p *is a prime.*

Proof. See [61] or [60].   □

8.2 Monodromy groups of polynomials

Let $f(x)$ be a polynomial of degree $n > 0$, and consider the solutions, $x_i(c)$, of the equation $f(x) = c$. Let S be the set of critical points $c \in \mathbb{C}$ for which $f(x) = c$ and $f'(x) = 0$ has a common solution. Clearly there are at most $n(n-1)$ of these points. As c takes values in $\mathbb{C} - S$ the functions $x_i(c)$ are well defined. The group $G = \pi_1(\mathbb{C} - S)$ acts on the $x_i(c)$. The action is always transitive if we consider large values of c.

Definition 8.5. Let G be as above, then the action of G on the set of the x_i is called the *monodromy* group of the polynomial f, denoted $\mathrm{Mon}(f)$.

As mentioned in Chapter 6, we have the following theorem [64].

Theorem 8.6. *The monodromy group is isomorphic to the Galois group of $f(x) - c$ considered as a polynomial over $\mathbb{C}(c)$.*

Definition 8.7. We say that a polynomial $f(x)$ is *decomposable* if and only if there exist two polynomials g and h, both of degree greater than 1, such that $f(x) = g(h(x))$.

Proposition 8.8. *Let f be a polynomial as above and let G be its monodromy group. Then:*

(i) *the action of G is imprimitive if and only if the polynomial f is decomposable;*

(ii) *the action is 2-transitive if and only if the divided differences polynomial $H(x, y) = (f(x) - f(y))/(x - y)$ is irreducible.*

We note that quantities of the form $(f(x) - f(y))/(x - y)$ were also the ones appearing in Chapter 3.

Proof. (i) It is a consequence of Definition 8.3 that G is imprimitive if and only if and only if there is an element x_i and group H lying strictly between G_{x_i}, the subgroup of G fixing x_i and G. Via the Galois correspondence, this gives an intermediate field K strictly between $\mathbb{C}(x_i)$ and $\mathbb{C}(c)$. By Lüroth's theorem, $K = \mathbb{C}(r(x_i))$ for some rational function r, and hence $c = s(r(x_i))$ for some rational function s. As in Chapter 4, we can show that r and s can be chosen to be polynomial, giving a factorization of f. The argument can clearly be reversed for the other implication.

(ii) Let x_i be a root of $f(x) = c$. Clearly $H(x, x_i) = 0$ for any other root $x = x_j$. G is 2-transitive if and only if there is an automorphism which fixes x_i and sends x_j to any other element x_k, $k \neq i$ (here we use the transitivity of G). This is the case if and only if H is irreducible over $\mathbb{C}(x_i)$. $\qquad\square$

Definition 8.9. The unique polynomial $T_n(x)$ which satisfies $T_n(\cos(\theta)) = \cos(n\theta)$ is called the *Chebyshev* polynomial of degree n. Equivalently $T_n((z + z^{-1})/2) = (z^n + z^{-n})/2$.

From the definition, the Chebyshev polynomial T_n has $n-1$ distinct turning points when $T_n = \pm 1$. Hence

$$T_n'^2 \mid (T_n(x)^2 - 1).$$

Conversely, any polynomial with this property is equivalent to a Chebyshev polynomial after composing with a linear function.

Theorem 8.10. *Let $f(x)$ be a polynomial of degree n and $G = \mathrm{Mon}(f)$, then one of the following holds.*

(i) *The action of G on the x_i is 2-transitive.*

(ii) *The action of G on the x_i is imprimitive.*

(iii) *f is equivalent to a Chebyshev polynomial T_p where p is prime.*

(iv) *f is equivalent to x^p where p is prime.*

Remark 8.11. In particular, the question of whether f is a composite polynomial or not, can be solved very simply by considering whether or not the divided differences polynomial factorizes or not, having excluded the two exceptional cases above. "Equivalence" refers to pre- and post- composition by linear functions.

Proof. When c is large the x_i can be expanded as

$$x_i = \omega^r c^{1/n} + O(c^{(1/n)-1}),$$

where ω is an n-th root of unity. Thus, taking a sufficiently large loop in $\mathbb{C} - C$, we obtain an element of G which is an n-cycle. This element generates a subgroup $\mathbb{Z}_n$ of G which acts regularly on the roots of $f = c$.

Thus we can apply the Burnside–Schur Theorem above to show that the group must be 2-transitive, imprimitive, or a subgroup of $\mathrm{Aff}(\mathbb{Z}_p)$. In the latter case we note that every element of $\mathrm{Aff}(\mathbb{Z}_p)$ fixes at most one element of $\mathbb{Z}_p$. This means that for every critical value of f there is at most one x_i that remains fixed as we turn around this value.

Now, suppose f has r distinct critical values, $c_1, \ldots, c_r$, and f has r_i distinct turning points associated to the critical value c_i. Let the multiplicities of the roots of f' at these turning points be $m_{i,1}, \ldots, m_{i,r_i}$. Since a root of multiplicity $m_{i,j}$ gives a cycle of order $m+1$, then for all i we must have

$$n - 1 \leq \sum_{j=1}^{r_i} (m_{i,j} + 1) \leq n, \tag{8.2}$$

since at most one of the x_i remains fixed turning around each critical value. Summing these equations over i we obtain

$$r(n-1) \leq \sum_{i=1}^{r} \sum_{j=1}^{r_i} (m_{i,j} + 1) \leq rn. \tag{8.3}$$

But the number of turning points of f counted with multplicity is just the sum of the $m_{i,j}$, and hence

$$r(n-1) \le (n-1) + \sum_{i=1}^{r} r_i \le rn. \tag{8.4}$$

Since the sum of the r_i is at most $n - 1$ we must have $r \le 2$.

If $r = 1$, then (8.4) shows that $r_1 = 1$, and therefore $f(x)$ must have a root of multiplicity n. This is just Case (iv).

If $r = 2$ we need $n - 1 \le r_1 + r_2 \le n + 1$. But since r_i can be no more than $n/2$ this means that both r_i lie between $(n-1)/2$ and $n/2$. This implies that every turning point must have multiplicity 1 and the polynomial must be Chebyshev. $\qquad\qquad\square$

8.3 Proof of the theorem

We consider the level curves of the Hamiltonian $H = y^2 - f(x) = c$ as a two-sheeted covering of the complex plane $\mathbb{C}$ given by projection onto the x-axis. The sheets ramify at the roots of $f(x) = c$. Taking S to be the set of critical points as above, we let c vary in $\mathbb{C} - S$, and follow the effect on the homology group $H_1(\phi_c, \mathbb{Z})$. We wish to relate this group to the monodromy group of the polynomial $f(x)$. As x tends to infinity along the positive real axis, we can distinguish the two sheets as "upper" and "lower" depending on whether $y = \pm x^{n/2}$. We let t denote the deck transformation which takes y to $-y$ fixing x.

Let $H_1^c(\phi_c, \mathbb{Z})$ represent the closed homology group of ϕ_c over $\mathbb{Z}$. This can be obtained from $H_1(\phi_c, \mathbb{Z})$ by adding curves starting and finishing at infinity. Let $x_i(c)$ be the roots of $f(x) = c$. Generically, the x_i will have distinct imaginary parts, and so any closed path in $\mathbb{C} - S$ can be deformed so that only two of the x_i's have the same imaginary part at the same time. In other words, we can decompose every element of $\mathrm{Mon}(f)$ as a number of swaps of x_i's with neighboring real values.

Suppose that the x_i are initially numbered in order of decreasing imaginary part for a value of c close to zero. We let L_i represent the path from infinity (from the direction of the positive real axis) on the upper sheet, turning around x_i in the positive direction and returning to infinity on the lower sheet. Clearly $t(L_i) + L_i$ is homotopic to zero, and so the L_i generate $H_1^c(\phi_c, \mathbb{Z})$. Furthermore, the elements $L_i - L_{i+1}$ generate $H_1(\phi_c, \mathbb{Z})$.

The effect of a swap of x_i and x_{i+1} is to take L_{i+1} to L_i and L_i to $2L_i - L_{i+1}$. This is a little too complex to analyze in general, except for very specific systems. Instead we shall work for the moment over $\mathbb{Z}_2$. That is, we consider the images of the L_i in $H_1^c(\phi_c, \mathbb{Z}_2)$ and $H_1^c(\phi_c, \mathbb{Z}_2)$.

Working modulo 2 means that a swap of x_i and x_{i+1} takes L_{i+1} to L_i and L_i to L_{i+1}. That is, the action of $\mathrm{Mon}(f)$ on the L_i (mod 2) is exactly the same as the action on the x_i.

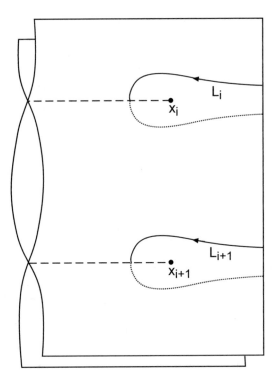

Figure 8.1: The loops L_i.

We now apply the results of Theorem 8.10 in order to prove Theorem 8.1. According to Theorem 8.10 we only need to consider four cases. The last two of these are easy. In the case that (iii) holds, we trivially have Case (iii) of Theorem 8.1, and in Case (iv) the Hamiltonian does not have a Morse point, and hence no tangential center. We shall show that the cases (i) and (ii) of Theorem 8.10 correspond to Cases (i) and (ii) of Theorem 8.1, and the proof is complete.

Case (i). If the monodromy group of f is 2-transitive, then we can find a transformation which takes any two x_i's to any other two. Since, working modulo 2, the action on the loops L_i is the same as the action on the x_i, we can find an element of the monodromy group which takes $L_i - L_{i+1}$ to $L_j - L_{j+1}$ modulo 2 for all i and j.

Now, the vanishing cycle γ_c occurs at the coalescence of two of these x_i's and so must correspond to one of the $L_k - L_{k+1}$ for some k. Thus, there exist paths

ℓ_i in $\mathbb{C} - C$ such that

$$\sigma(\ell_i)\gamma_c = L_i - L_{i+1} \pmod 2,$$

for all i.

Now let $N = 2\lfloor (n-1)/2 \rfloor$. Then $L_i - L_{i+1}$ form a basis of $H_1(\phi_c, \mathbb{Z})$. From the discussion above, we have

$$\begin{pmatrix} \sigma(\ell_1)\gamma_c \\ \sigma(\ell_2)\gamma_c \\ \vdots \\ \sigma(\ell_N)\gamma_c \end{pmatrix} = A \begin{pmatrix} L_1 - L_2 \\ L_2 - L_3 \\ \vdots \\ L_N - L_{N+1} \end{pmatrix},$$

where the matrix A reduces to the identity matrix if we reduce modulo 2. In particular, A is invertible, and we can express the basis of $H_1(\phi_c, \mathbb{Z})$ as sums of the $\sigma(\ell_i)\gamma_c$ with coefficients in $\mathbb{Q}$. That is, γ_c generates $H_1(\phi_c, \mathbb{Q})$. This gives us Case (i) of Theorem 8.1.

Case (ii). In this case $\mathrm{Mon}(f)$ is imprimitive (but nevertheless transitive) on the set of roots, $S = \{x_1, \ldots, x_n\}$. Let S_1 be one of the subsets in the decomposition $S = \bigcup S_i$, and let $s = x_k$ be an element of S_1. We denote G_s and H the subgroups of $G = \mathrm{Mon}(f)$ which leave s and S_1 fixed respectively. Then Case (iii) implies that

$$G_s \subsetneqq H \subsetneqq G. \tag{8.5}$$

As stated in Section 8.2, the group $G = \mathrm{Mon}(f)$ is just the Galois group of $\mathbb{C}(x_1(c), \ldots, x_n(c))$ over $\mathbb{C}(c)$. We consider the corresponding fixed fields of the groups in (8.5) under the Galois correspondence, to obtain

$$\mathbb{C}(x_k(c)) \supsetneqq K \supsetneqq \mathbb{C}(c), \tag{8.6}$$

where K is the fixed field of H. From Lüroth's theorem, we must have $K = \mathbb{C}(r(x_k))$, for some rational function $r(x_k) \in \mathbb{C}(c_k)$. Then (8.6) implies that $c = s(r(x_k))$ for some rational function s. Thus $f(x) = s(r(x))$, and a similar argument to the one given in Lemma 4.6 shows that s and r can in fact be chosen to be polynomials.

This completes the proof of Theorem 8.1. $\qquad\square$

8.4 The symmety of the differential

We now wish to relate the results of Theorem 8.1 to the form of the perturbation term ω. Clearly, from the form of H, any perturbation term ω can be written in the form $\omega = dB' + A'\,dx$ for some choice of polynomials A' and B'. By the repeated use of the identity

$$2\xi'(x)y^{n+2}\,dx = (n+2)\xi(x)f'(x)y^n\,dx + d(2\xi(x)y^{n+2}) - (n+2)\xi(x)y^n\,dF,$$

for any choice of polynomial $\xi(x)$, we can reduce ω to the form:

$$\omega = A\,dH + dB + y\,g(x)\,dx,$$

for some choice of polynomials A and B. Since the first two terms on the right hand side are relatively exact, it is sufficient to consider the case where $\omega = y\,g(x)\,dx$. If we assume our vanishing cycle is at the origin, then we can choose $f(x) = x^2 + O(x^3)$ after a scaling. We choose a coordinate $X = X(x)$ such that

$$X(x)^2 = f(x), \qquad X(x) = x + O(x^2).$$

Then, we have that

$$\oint_{\gamma_c} y\,g(x)\,dx = \oint_{X^2+y^2=c} y\,m(X)\,dX,$$

where

$$m(X(x)) = \frac{2g(x)}{f'(x)}.$$

This integral will vanish if and only if $g(0) = 0$ and m is an even function of X. Thus we require

$$m(X(x)) = \phi(X(x)^2) = \phi(f(x)),$$

for some analytic function ϕ, and hence $2g(x) = f'(x)\,\phi(f(x))$.

   If we let $G(x)$ be the primitive of $g(x)$ with $G(0) = 0$, and Φ be the primitive of ϕ with $\Phi(0) = 0$, then we have $2G(x) = \Phi(f(x))$.

   If $f(x)$ is a composite polynomial, then both $G(x)$ and $f(x)$ have a common factor $h(x)$ (this can be shown in a similar manner to the arguments in Chapter 4), so that $f = s \circ h$ and $2G = r \circ h$ for some polynomials r and s which gives

$$\omega = y\,g(x)\,dx = y\,r'(h(x))\,h'(x)\,dx = \pi^*(y\,r'(z)\,dz).$$

   Thus the original integrand can be seen to be the sum of a relatively exact term and a part which is the pullback of an Abelian integral on the factored system with Hamiltonian $y^2 + r(x)$.

   If the push forward of the cycle to the factored system is homotopic to zero, then the original integral must vanish. However, if this is not the case, we can iterate the process for the integral on the factored system, which must also vanish. Since the pull back of a relatively exact term is also relatively exact, we see that the original integrand that either the pullback term is also relatively exact, or the cycle is homotopic to zero after a finite number of factorization steps.

Notes

§1 The monodromy group of a polynomial is an object of some interest in the inverse Galois problem (see, for example, the work of Müller [98], from whom I first learnt about the Burnside–Schur results).

Complete details of the above result together with an analysis of the Chebyshev case and other related results can be found in [54].

§2 One of the intersting features of the problem is that much of the proof revolves around the behaviour of the solutions of $f(x) = c$. There is a growing literature on "zero-dimensional abelian integrals", which examines the behaviour of zero dimensional cycles attached to the roots of a polynomial equation $f(x) = c$ in detail [2, 3, 69].

§3 The case of the Hamiltonian $H = y^4 + f(x)$ has been discussed by Lopez [87], who also considers the case $g(y) + f(x)$ for f and g of degree 4. It would be interesting to know more about the general case. In particular, whether this factorization property holds for other Hamiltonians.

Problem 8.1. If a vanishing cycle of a Hamiltonian system doesn't generate the homology of a generic level curve, does it imply that the Hamiltonian has a non-trivial factorization?

The above problem is topological only, but one could also ask about the integrand itself.

Problem 8.2. Suppose there is a non-trivial factorization of the Hamiltonian and an Abelian integral is identically zero on a vanishing cycle. Is the integrand a sum of a relatively exact term and a term which can be seen as a pull back of an Abelian integral on a factored Hamiltonian and where the push forward of the cycle is homotopic to zero?

Chapter 9

Holonomy and the Lotka–Volterra System

In this section we give another idea related to monodromy. This is the holonomy of the foliation $P\,dy - Q\,dx = 0$ associated to the system (1.2) in the neighborhood of an invariant curve. This object, roughly speaking, is the nonlinear analog to the monodromy of the solutions of a linear differential equation as they turn around a singular point. Alternatively, it can be thought of as a kind of Poincaré return map for foliations.

We shall define the holonomy in the next section and give a very basic theorem which will guarantee the integrability of a critical point. Recall this means the following:

Definition 9.1. The origin of the analytic system

$$\dot{x} = P(x,y) = x + \tilde{P}(x,y), \qquad \dot{y} = Q(x,y) = -\lambda y + \tilde{Q}(x,y), \qquad (9.1)$$

where $\tilde{P}$ and $\tilde{Q}$ contain terms of order 2 or higher, is *integrable* if there exists an analytic change of coordinates $(X,Y) = (x + o(x,y), y + o(x,y))$ in the neighborhood of the origin transforming the system into

$$\dot{X} = Xh(X,Y), \qquad \dot{Y} = -\lambda Y h(X,Y), \qquad (9.2)$$

where $h = 1 + O(X,Y)$. Alternatively, the origin of (9.1) is integrable if and only if there exist holomorphic functions $X = x + o(x,y)$ and $Y = y + o(x,y)$ such that $X^\lambda Y$ is a first integral of (9.1).

When $\lambda = p/q$ we call (9.1) a *p:-q saddle*. If (9.1) it is not linearizable, we call it a *resonant saddle*. The problem of finding whether a $p : -q$ saddle is integrable or not is in exact analogy to the center-focus problem. Indeed, if we assume that a transformation has straightened out the separatrices, so that $\tilde{P} = x\hat{P}$ and $\tilde{Q} = y\hat{Q}$, we can take $x = X^q$ and $y = Y^p$ we obtain a new system (after scaling time by a constant factor q).

$$\dot{x} = X + \hat{P}(X^q, Y^p), \qquad \dot{Y} = -Y + \hat{Q}(X^q, Y^p)q/p, \qquad (9.3)$$

© The Author(s), under exclusive license to Springer Nature Switzerland AG 2024
C. Christopher et al., *Limit Cycles of Differential Equations*, Advanced Courses
in Mathematics - CRM Barcelona, https://doi.org/10.1007/978-3-030-59656-9_9

and the origin of (9.1) is integrable if and only if the origin of (9.3) is integrable (i.e., it is complex center). We can therefore apply the same techniques of computing Lyapunov quantities etc., in order to detect whether the system (9.1) is integrable or not.

     The holonomy around the separatrix of a saddle turns out to be linearizable if and only if the saddle is integrable. Furthermore, if the separatrix is known, then we can relate this holonomy to the holonomy of the other critical points on the separatrix. Under favorable conditions, we can show that the original holonomy must be linearizable and hence give a potentially new method of finding integrable critical points.

     We apply our results to the origin of the Lotka–Volterra system

$$\dot{x} = x(1 + ax + by), \qquad \dot{y} = y(-\lambda + cx + dy), \tag{9.4}$$

and show that this technique seems to explain many of the cases of integrability when $\lambda = p/q$ for $p + q \leq 12$, apart from a small number of Darboux cases with additional invariant algebraic curves.

     These exceptional cases turn out, in fact, to be also explainable in terms of monodromy conditions, but now using the monodromy around the additional curve. Including this additional form of monodromy, it turns out we can explain the integrability of not just the critical point at the origin, but also the other critical points of the system. We conjecture that all integrability conditions are given by this method for the Lotka-Volterra systems. To what extent this is mirrored in other families of systems is an open problem.

     Further details can be found in [57], [75] and [42] from whence the material is drawn.

9.1   The monodromy group of a separatrix

In this section we consider a saddle point with a separatrix given by either an invariant line or a non-singular conic and give sufficient conditions for the integrability of a saddle point by looking at the monodromy group of the separatrix. We apply this to the Lotka–Volterra equations, to obtain four classes of explicit conditions which give integrable critical points.

     The surprising thing is that, even though these conditions on the monodromy groups are elementary, they comprise all the known cases of integrability for the Lotka–Volterra equations, except for the case where the system has an invariant straight line and two exceptional Darboux integrable cases [27, 96]. In the latter two cases, we can extend the monodromy method to include the monodromy around this invariant curve (see later).

     Consider the foliation on $\mathbb{CP}^2$ generated by the 1-form associated to the vector field. Let Γ be an invariant line or conic for the 1-form, and $Q_1, \ldots, Q_n$ be the singular points of the foliation which lie on Γ. For (9.4) we have three such lines: the two axes and the line at infinity. Clearly $\Gamma' = \Gamma - \{Q_1, \ldots, Q_n\}$ is isomorphic to an n-punctured sphere.

Choose a family of analytic transversals, Σ_x, through each point x in Γ', and fix a base point, P, in Γ', and an analytic parameterization z of Σ_P with $z = 0$ corresponding to the point P. For each path γ in $\pi(\Gamma', P)$, we can define a map from a neighborhood of P in Σ_P to Σ_P by lifting the path γ to the leaf of the foliation though $s \in \Sigma_P$ via the transversals Σ_x, $x \in \gamma$. Using the parameter z, this map can be identified with the germ of a diffeomorphism from $\mathbb{C}$ to itself, fixing $z = 0$. We call the set of all such diffeomorphisms $\mathrm{Diff}(\mathbb{C}, 0)$.

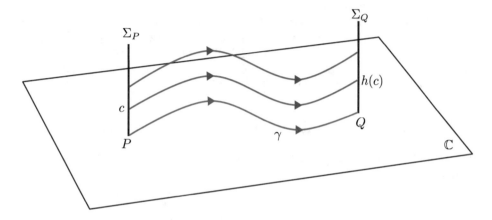

Figure 9.1: Lifting a curve γ to obtain a map from Σ_P to Σ_Q.

Clearly the map $M : \pi(\Gamma', P) \to \mathrm{Diff}(\mathbb{C}, 0)$ is in fact a group homomorphism. We denote the image of the path γ by M_γ. The *monodromy group* is the image of M. The *monodromy of one singular point* Q_i is M_γ where γ is a loop turning around Q_i exactly once in the positive direction and not containing any other singular point in its interior.

The map M_γ depends only on the homotopy type of γ in Γ'. If we use a different base point P_1, then the two monodromy groups are conjugate. Likewise a different choice of transversals and their parameterizations has the effect of conjugating the group. Thus the following notions for the monodromy of a singular point are intrinsic:

- the monodromy of the singular point is the identity;

- the monodromy of the singular point is linearizable.

Theorem 9.2. *Consider a polynomial system with a saddle point at the origin*

$$\begin{aligned} \dot{x} &= x(1 + P(x, y)) = x(1 + O(x, y)), \\ \dot{y} &= -\lambda y + Q(x, y) = -\lambda y + o(x, y), \end{aligned}$$

$$(9.5)$$

where $\lambda > 0$. If all singular points of the system on the y-axis (including any critical points on the line at infinity) except the origin are integrable and if all of them but one have identity monodromy maps corresponding to the invariant y-axis, then the origin is also integrable.

Proof. We consider the completion of the line $x = 0$ as the Riemann sphere S^1. Let $Q_1, \ldots, Q_n$ be the singular points of the system on that leaf.

Let Q_i be a point of saddle or node type. It is known that Q_i is integrable if and only if the corresponding monodromy map is linearizable (this is proved in [95] and [105] for a saddle. For a node it can easily be proved by considering the analytic normal form at the node).

Take a base point $y_0 \in S^1 - \{Q_1, \ldots, Q_n\}$ and loops γ_i from y_0 winding once around the singular points Q_i in the positive sense; then γ_1 is homotopic to $\gamma_n^{-1} \circ \cdots \circ \gamma_2^{-1}$, with appropriate re-labelling of the Q_i. As a result M_{γ_1} is conjugate to $M_{\gamma_n}^{-1} \circ \cdots \circ M_{\gamma_2}^{-1}$. Since all of them are the identity except one which is linearizable, then the map M_{γ_1} is linearizable. $\qquad\square$

9.2   Integrable points in Lokta–Volterra systems

We apply these results to the Lotka–Volterra family (9.4). This family is invariant under

$$(x, y, t, \lambda, a, b, c, d) \mapsto \left(-\lambda y, -\lambda x, -\frac{t}{\lambda}, \frac{1}{\lambda}, d, c, b, a\right) \qquad (9.6)$$

and corresponding cases under this invariance are called *dual*.

Lemma 9.3. *A node is linearizable if and only it it has two analytic separatrices.*

Proof. A node with eigenvalues λ_1, λ_2 whose quotient is in $\mathbb{R}^+$ can always be brought to normal form by an analytic change of coordinates. When $\frac{\lambda_2}{\lambda_1} \notin \mathbb{N} \cup 1/\mathbb{N}$, then the normal form is linear and the two axes are analytic separatrices. When $\frac{\lambda_2}{\lambda_1} = n \in \mathbb{N}$ the normal form is

$$\begin{aligned} \dot{x} &= \lambda_1 x, \\ \dot{y} &= \lambda_2 y + \alpha x^n. \end{aligned} \qquad (9.7)$$

If $\alpha = 0$, then the system is linear as before and all integral curves through the origin are analytic, while if $\alpha \neq 0$ the curve $x = 0$ is the unique analytic integral curve through the origin. Similarly for $\frac{\lambda_2}{\lambda_1} \in 1/\mathbb{N}$. $\qquad\square$

Theorem 9.4. *We consider the Lotka–Volterra system (9.4) with $\lambda > 0$. Then the origin is integrable if one of the following conditions is satisfied.*

(A_n). $\lambda + \frac{c}{a} = n$ *with* $n \in \mathbb{N}$, $2 \leq n < \lambda + 1$.

(B_n). $\frac{b}{d} + \frac{1}{\lambda} = n$ *with* $n \in \mathbb{N}$, $2 \leq n < \frac{1}{\lambda} + 1$.

(C_n). $\frac{c}{a} + n = 0$ *with* $n \in \mathbb{N} \cup \{0\}$ *and* $n < \lambda$ *and* $\lambda \neq n + \frac{1}{m}$ *with* $m \in \mathbb{N}$.
        If $\lambda = n + \frac{1}{m}$, *then an additional condition is necessary for integrability.*

(D_n). $\frac{b}{d} + n = 0$ with $n \in \mathbb{N} \cup \{0\}$ and $n < \frac{1}{\lambda}$ and $\frac{1}{\lambda} \neq n + \frac{1}{m}$ with $m \in \mathbb{N}$.
If $\frac{1}{\lambda} = n + \frac{1}{m}$, then an additional condition is necessary for integrability.

$(E_{n,m})$. $\lambda + \frac{c}{a} = n$ and $1 - \frac{b}{d} = \frac{1}{m}$ with $n, m \in \mathbb{N}$, $n > 1$ and $0 < \frac{(c-a)(d-b)}{ad-bc} \notin \mathbb{N}$.

$(F_{n,m})$. $\frac{1}{\lambda} + \frac{b}{d} = n$ and $1 - \frac{c}{a} = \frac{1}{m}$ with $n, m \in \mathbb{N}$, $n > 1$ and $0 < \frac{(c-a)(d-b)}{ad-bc} \notin \mathbb{N}$.

$(G_{n,m})$. $\lambda + \frac{c}{a} = n$, $1 - \frac{b}{d} > 0$ and $\frac{ad-bc}{(c-a)(d-b)} = m$ with $m, n \in \mathbb{N} - \{1\}$.

$(H_{n,m})$. $\frac{1}{\lambda} + \frac{b}{d} = n$, $1 - \frac{c}{a} > 0$ and $\frac{ad-bc}{(c-a)(d-b)} = m$ with $n, m \in \mathbb{N} - \{1\}$.

(Note that some strata with different names may be identical for some values of λ and of the indices. This can for instance happen with $(E_{n,m})$ and $(G_{n,m'})$.)

Proof. To apply the previous theorem and corollary we need to calculate the Jacobian matrix and the eigenvalues at all singular points along the axes and along infinity. On each separatrix there are three critical points: the one at the origin with ratio of eigenvalues $-\lambda$, one in the finite plane, and one where the axes cross the line at infinity. The Jacobians for the finite critical points $P_1 = (-\frac{1}{a}, 0)$ (resp. $P_2 = (0, \frac{\lambda}{d})$) on the x-axis (resp. y-axis) are

$$\begin{pmatrix} -1 & -\frac{b}{a} \\ 0 & -\lambda - \frac{c}{a} \end{pmatrix} \quad resp. \quad \begin{pmatrix} 1 + \lambda\frac{b}{d} & 0 \\ \lambda\frac{c}{d} & \lambda \end{pmatrix}, \tag{9.8}$$

showing that the monodromy of the finite critical points on the x-axis (resp. y-axis) is the identity if $\lambda + \frac{c}{a} = n$ (resp. $\frac{b}{d} + \frac{1}{\lambda} = n$) with $n \in \mathbb{N}$, $n \geq 2$.

We now study the singular points at infinity. For that purpose we first consider the chart $(u, z) = (y/x, 1/x)$ to calculate the Jacobian matrix at the intersection of the line at infinity with the x-axis, which we denote $P_x = (0, 0)$. We can also calculate the Jacobian at the other critical point $P_\infty = (\frac{a-c}{d-b}, 0)$ on the line at infinity. After multiplication by z, the system becomes:

$$\begin{aligned} \dot{u} &= (c - a)u + (d - b)u^2 - (1 + \lambda)uz, \\ \dot{z} &= -az - buz - z^2, \end{aligned} \tag{9.9}$$

yielding the following Jacobian matrices for P_x and P_∞:

$$\begin{pmatrix} c - a & * \\ 0 & -a \end{pmatrix} \quad resp. \quad \begin{pmatrix} -(c - a) & * \\ 0 & \frac{ad-bc}{b-d} \end{pmatrix}. \tag{9.10}$$

Similarly the chart $(v, w) = (x/y, 1/y)$ is used to study the infinite singular point P_y along the y-axis. Its Jacobian matrix is given by

$$\begin{pmatrix} b - d & * \\ 0 & -d \end{pmatrix}. \tag{9.11}$$

We can represent the ratios of eigenvalues on the diagram below, where the arrows represent the direction of the eigenvalue which is the numerator of the eigenvalue ratio.

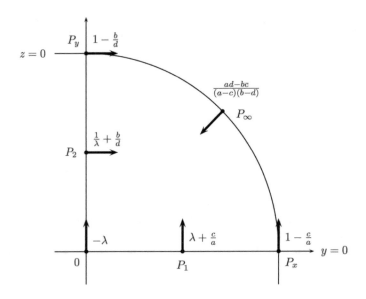

Figure 9.2: Ratio of eigenvalues for the Lotka–Volterra system

Note that the sum of the eigenvalue ratios along any line is equal to 1. This follows from the index formula of Lins Neto [84]. More generally, for a non-singular curve of degree n, the sum of the ratio of eigenvalues is equal to n^2, the self-intersection number of the curve.

We now prove the cases (A)–(H) given above. We may remove the indices when they are not necessary.

CASE $(A_n)/(B_n)$: In Case (A_n), the condition implies that the monodromy of P_1 corresponding to the invariant x-axis is the identity and the critical point P_x is a node. It is always linearizable since there are two analytic separatrices. Case (B_n) is the dual of Case (A_n).

CASE $(C_n)/(D_n)$: Case (C_n) is similar to Case (A), but now the monodromy at P_x is the identity corresponding to the invariant x-axis, and P_1 is a node. It is linearizable if $\lambda \neq n + \frac{1}{m}$ with $m \in \mathbb{N}$, otherwise (the case of a resonant node) the obstruction to linearizability consists of only one condition. Case (D_n) is the dual of Case (C_n).

CASE $(E_{n,m})/(F_{n,m})$: Case $(E_{n,m})$ requires a double application of Theorem 9.2. The conditions imply that the monodromy of P_1 corresponding to the invariant x-axis is the identity. Thus the monodromy at the origin is conjugate to the inverse of the monodromy of P_x corresponding to the invariant x-axis. Now, this monodromy is linearizable if and only if P_x is integrable. This is the case if and only if the monodromy of P_x corresponding to the other separatrix (in this

- case, the line at infinity) is linearizable. Now, the conditions given in Case $(E_{n,m})$ guarantee that the monodromy of P_y corresponding to the line at infinity is the identity. (P_y is a node with ratio of eigenvalues $m \in N$.) Hence the origin is integrable if and only if the monodromy of P_∞ is linearizable corresponding to the line at infinity. Now the final condition in Case $(E_{n,m})$ guarantees that P_∞ is a non-resonant node, and therefore linearizable. Case $(F_{n,m})$ is the dual of Case $(E_{n,m})$.

CASE $(G_{n,m})/(H_{n,m})$: Case $(G_{n,m})$ is the same as Case (E) except that now, the monodromy of P_∞ corresponding to the line at infinity is the identity and the point P_y is a node (necessarily linearizable). Case $(H_{n,m})$ is the dual of Case $(G_{n,m})$. □

We have the following conjecture.

Conjecture 9.5. The Lotka–Volterra system (9.4) with $\lambda \in \mathbb{Q}^+$ is integrable if and only if either

1. the system has a third invariant line, i.e.,

$$\lambda ab + (1 - \lambda)ad - cd = 0; \tag{9.12}$$

2. one of the conditions of Theorem 9.4 is satisfied;

3. or there is an invariant algebraic curve, $f = 0$.

The first and third items above will give Darboux centers. In fact, we know from the lists given in [27, 96] that there are essentially only two cases (with $\lambda = 8/7$ and $13/7$ and their duals) where this last condition holds, which are not contained in the previous two conditions. These (after scaling) are the systems

$$\begin{aligned} \dot{x} &= x(1 - 2x + y), \\ \dot{y} &= y(-\tfrac{8}{7} + 4x + y), \end{aligned} \tag{9.13}$$

with invariant cubic

$$F(x, y) = 1372xy(3x - y) - 1764xy - 63y - 72 = 0, \tag{9.14}$$

and

$$\begin{aligned} \dot{x} &= x(1 - 2x + y), \\ \dot{y} &= y(-\tfrac{13}{7} + 4x + y), \end{aligned} \tag{9.15}$$

with the invariant quartic

$$F(x, y) = 343x^2 y(3x - y) - 588x^2 y + 21xy + 18x - 9 = 0, \tag{9.16}$$

together with their duals.

Theorem 9.6. *The conjecture is proved for* $\lambda = \frac{p}{q}$ *with* $p + q \le 12$ *and all* $\lambda = \frac{n}{2}$ *and* $\lambda = \frac{2}{n}$ *for* $n \in \mathbb{N}$.

Proof. The proof consists in calculating the Lyapunov quantities for the origin of •
(9.4) and finding their common roots. As many as three quantities are needed to
give complete conditions. The conditions for $\lambda = 1/n$ and $2/n$ for $n \in \mathbb{N}$ were given
in [67] and [73] respectively. In these cases we can prove by a counting argument
that the list of conditions is necessary and sufficient: it is easy to prove that
the first two saddle quantities cannot vanish elsewhere than the known sufficient
conditions. Independent calculations of some of these cases have been done in
[85].   □

9.3 Holonomy on more general curves

The holonomy method can also be applied in the neighbourhood of a curve of
higher degree. Clearly, if the curve is non-singular and has genus zero, then the
conditions given in Theorem 9.2 will still imply the existence of an integrable
critical point. If the genus, g, is greater than zero, however, additional conditions
will have to be imposed to ensure that the holonomy around the $2g$ non-trivial
loops on the curve is trivial. As this is a non-local problem, verifying this condition
is unlikely to be easy. We are therefore reduced to consider curves of genus 0, and
hence of degree 1 and 2 only.

On the other hand, if the curve is singular, we could have curves of higher
degree of genus zero. The holonomy can still be calculated in these cases by blowing
up the critical point. A-priori, this might seem quite complex, but it turns out that
in most cases it is easy to show that the holonomy is linearizable since singularities
like cusps can only pass through a node if the critical point is non-degenerate.

It turns out that for the two exceptional cases, (9.13) and (9.15), the additional invariant curves allow a monodromy argument to be applied.

More generally, it is possible to generalize the calculations above for Lotka-
Volterra systems to classify all $p : -q$ resonant critical points, not just the one at
the origin. The monodromy arguments can then be extended to cover the critical
points at the origin with $p + q \leq 20$, or those on an axis with $p + q \leq 17$, or the
unique critical point not lying on the axes (the hardest case), with $p+q \leq 10$ [42].
In each case there is either an invariant line (that is, equation (9.12) holds) and
the system is Darboux integrable, or there is a monodromy argument using the
invariant lines and curves of the system to show integrability.

A detailed examination of the possible invariant curves of the Lotka-Volterra
equation, based on the classification of Moulin Olagnier [96], is given in [75]. We
finish with the following problems.

Problem 9.1. Is is possible to relate the type of mondromy argument to other
integrability mechanisms (eg. Darboux integrals, symmetries, or blow down to a
node)?

Problem 9.2. Show that the monodromy conditions in Conjecture 9.5 are necessary. Is the signifance of monodromy techniques unique to Lotka Volterra systems?

Notes

§1 It is not clear whether the cases proved to be a center by these monodromy arguments comprise new types of centers or whether they can be related to the other known mechanisms (hence Problem 9.1 above). A more detailed examination of the Lotka–Volterra system can be found in [57].

It would be also interesting to know if there can be non-trivial real centers given by monodromy arguments.

There are many applications of holonomy techniques to the whole area of integrability of polynomial vector fields. In particular, there is a nice dictionary which, in generic cases, matches Liouvillian solutions in a neighborhood of an invariant algebraic curve to the solvability of its monodromy group.

The general analytic classification of resonant saddles and their holonomy has been done by Martinet and Ramis [94], however we do not need to use this work here, as we are only interested in integrable saddles and these have only linear holonomy maps. The classification of saddles for irrational λ is much more complex and is not known in general.

§2 For three dimensional Lotka-Volterra systems (1.24), a similar consideration still holds. In this case, the holonomy can be calculated in the neighborhood of an invariant line or curve, but will now be expressed as a map from $(\mathbb{C}^2, 0)$ to itself. The adaption of the method of Theorem 9.2 requires a little more care. It is still the case that integrable points will have a linearizable monodromy, about that point, but the reverse implication will only hold if the ratio of eigenvalues in both directions away from the curve is negative, so that the leaves of the foliation can be extended to a full neighbourhood of the critical point.

Full classifications of the integrability conditions (i.e. two independent local first integrals) are known for a $(\lambda : \mu : \nu)$-resonant critical point at the origin in the following cases among others: $(1 : -1 : 1)$, $(2 : -1 : 1)$, $(1 : -2 : 1)$ [9]; $(i : -i, \lambda)$, $(i - 1, -i - 2, 2)$ [10]; $(1 : -3 : 1)$ [8]; $(3 : -1 : 2)$ [11]. The techniques required to prove integrability are quite varied: Darboux integrability; blow down to a critical point in the Poincaré domain; and various adhoc arguments constructing power series in one or two of the variables.

However, application of the monodromy method, with the modification indicated above, unifies the proof of most of these integrability conditions. Of the four systems considered in [9, 11], about half of the integrability cases yield directly to monodromy calculations. However, of the remaining systems, most seem to have a projection onto a simpler two dimensional system. In this case, the restriction on the ratio of eigenvalues given above can be relaxed and the monodromy method applied.

One new feature which is intriguing, is the frequent presence of a line of singularities when the integrability conditions are applied. Unlike the two dimensional case, such a line of singularities cannot be removed by a non-linear time scaling.

However, in the cases of interest, the ratio of of the two non-zero eigenvalues is constant along the line and is equal to 1 or 2. A single or double blow up takes the line of singularities to a plane of singularities which can now be removed to leave a non-singular point where it intersected another curve. Since the topology of the space outside the blow up locus is not altered by a blow up, the original monodromy at that point must be trivial. Thus, if a curve intersects such a line of singularities, the contribution to the monodromy on the curve at that singular point is trivial.

Problem 9.3. Investigate these problems further. Could monodromy methods also be applied to finding just one first integral in the case when the critical point is not completely integrable?

Chapter 10

Other Approaches

In this final chapter I want to mention briefly three other approaches to the general center-focus problem. In the first, we try to identify whole components of the center variety by finding their intersections with specific subsets of parameter space and then showing that the type of center is "rigid". In the second approach, we try to see the consequences of a center on its bounding separatrix cycle. Monodromy-type arguments play an implicit role in both of these approaches. The last section describes attempts to circumvent some of the computational complexity of the center focus problem by working over finite fields. It ends with an experimental approach to the center-focus via intensive computations using modular arithmetic and an application of the Weil conjectures. It makes a fitting conclusion to our range of monodromy techniques, since the arithmetic analog of monodromy was an essential ingredient in Deligne's proof of the Weil conjectures [59].

10.1   Finding components of the center variety

In order to be able to use results from algebraic geometry, we shall consider the center-focus problem for systems (1.2) with complex coefficients. We shall also compactify the space of parameters to $\mathbb{CP}(N)$ for some N. This can be done, for example, by taking the space of coefficients of P and Q in (1.2) modulo the action of $\mathbb{C}^*$ obtained by multiplying all coefficients by a constant (which of course does not affect the existence of a center).

Now, the closure of the center variety therefore becomes an algebraic subset of $\mathbb{CP}(n)$, which we denote Σ. Suppose Σ_1 is an irreducible component of Σ of dimension r, and let H be a subspace of $\mathbb{CP}(n)$ of dimension s; then if $r + s > n$ the two spaces must intersect in a non-empty space of dimension $r + s - n$.

In particular, if we restrict our attention to H, we can see a trace of all components of the center variety whose dimension is $n - s$ or greater. However, it might be possible that more than one component intersects H in the same subset, and so we cannot distinguish between the different components of Σ by just looking at H.

In order to be able to improve this result we need to be able to show that certain types of centers are "rigid". That is, if we know that at a certain parameter value we have this type of center, then all centers lying close to this value will also have a center of this type.

The most general result in this direction was found by Movasati [97]. In his paper, he shows that for a polynomial system (1.2) of degree d, the subset $\Sigma(d_1, \ldots, d_r)$ of the center variety which is composed of Darboux centers given by the curves $f_i = 0$ of degree d_i, $i = 1, \ldots, r$ with $d + 1 = \sum d_i$ and their limits is a full component of the center variety. Thus any point in H which has a center which is a generic point of $\Sigma(d_1, \ldots, d_r)$ cannot also be in the intersection of H with any other component of the center variety.

The proof is based on a reduction to exactly the case in Chapter 7 where the Hamiltonian has $d+1$ invariant lines in general position. The author then examines the tangent space of the center variety at that point using similar arguments to the ones given there, but including higher degree perturbations. The conclusion is that the tangent cone of Σ at this point is exactly spanned by the spaces $\Sigma(d_1, \ldots, d_r)$ as the d_i run over all subsets of integers summing to $d+1$. This is enough to show that the $\Sigma(d_1, \ldots, d_r)$ are complete components of Σ.

Unfortunately, this method does not give a method to determine whether a specific Darboux center is generic or not. The direct calculation of Darboux center perturbations [55], outlined in the notes to Chapter 7, confirms Movasati's conclusions, but now gives explicit conditions to guarantee that a Darboux center must be generic.

Problem 10.1. Develop the methods above to work with other Darboux components of the center variety where $\sum d_i > d + 1$.

This is a difficult problem since the set of invariant curves $f_i = 0$ are no longer generic and extra conditions need to be imposed on them in order to generate a differential system of degree d. It is not clear how to enumerate these or put them in a compact form, although many cases are now known for cubic systems [132].

A similar idea has been suggested in [24] for Abel systems (6.1). Here the authors consider the limit of families of centers of the form (6.3) as ϵ tends to infinity. Alternatively, after a rescaling, we could consider the systems

$$\frac{dy}{dx} = \epsilon p(x)y^2 + q(x)y^3, \qquad 0 \le x \le 1, \qquad (10.1)$$

as ϵ tends to zero. The situation is slightly more subtle than the one we have described above, as the limiting case always has a center. Thus the authors take the "tangential" part of the center conditions in order to define a "center at infinity". They then show that these centers must be solutions to the moment problem (6.4), which give centers of (6.3) for all ϵ. Every component of the variety of centers at infinity of dimension n must therefore extend to a component of the center variety of (6.3) of dimension $n + 1$. Some more work is needed to show that this is the unique component which intersects in this way with the centers at infinity. Conversely, every component of the center variety of dimension $n + 1$ must correspond to a component of the variety of centers at infinity of dimension n. Thus, the only centers which do not satisfy (6.4) exist only for discrete parameter values.

10.2 Extending Centers

A second approach is to analyze what happens at the boundary of a period annulus. Our hope is that the effect of a separatrix cycle having an identity return map has strong global consequences for the system. We could even hope that the local first integral of the center could be extended, in some ramified way, past these boundaries to obtain further global consequences of a center.

In general this seems a very difficult problem, but in one case this problem is quite easy. This is when the boundary of the center is a homoclinic loop attached to a saddle.

The nature of the return map in the neighborhood of a saddle in this case is well known [80, 113, 114]. In particular, the asymptotics are governed by two sets of interleaving terms. One set comes from the loop and the others are essentially governed by the Lyapunov quantities of the saddle. In order for the return map for a homoclinic loop to be the identity (or even analytic), we need to have all the Lyapunov quantities vanish, and therefore we can conclude that the saddle is integrable.

Thus in this case, the local integrability of the center has a global effect on a neighboring critical point. Furthermore, since the saddle is integrable, the local first integral of the center can be extended beyond the boundary of the homoclinic loop and hopefully could give further information about the system.

Example 10.1. We give an application of this idea to the problem of the center for the Abel equations (6.1). We assume that $q(x)$ does not vanish at $x = a$ or $x = b$, and has at most one root in (a, b) with $q' < 0$ at this root, and show that the system must then satisfy the composition condition of Chapter 6. We take $a = 0$ and $b = 1$ as before.

In order to prove this result, we note that the transformation $z = 1/y$ brings (6.1) to the form

$$\dot{x} = z, \qquad \dot{z} = -q - zp. \tag{10.2}$$

If $y(x, c)$ is the solution of (6.1), the solution of (10.2) is given by $z(x, c) = 1/y(x, 1/c)$ with $z(0, c) = c$ for c sufficiently large. Thus, if there is a center for (6.1), then there is a band of trajectories of (10.2) which passes between $x = 0$ and $x = 1$ with $z(1, c) = z(0, c)$ for c sufficiently large. We shall first show that the boundary of this band of trajectories must consist of a non-degenerate saddle and two of its separatrices (see Figure 10.1)

Consider the path of the trajectory $z(x, c)$ as c decreases. Elementary considerations show that such a trajectory which crosses the line $z = 0$ in $(0, 1)$ cannot

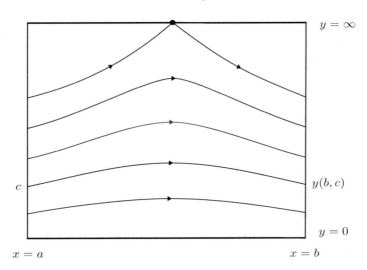

Figure 10.1: The return map extended to infinty.

pass from $x = 0$ to $x = 1$. On the other hand, since all the critical points of the system lie on the line $z = 0$, we can certainly decrease c until $z(x, c)$ impinges on $z = 0$ at some point. Clearly there are only two possibilities: either we can decrease c to zero, with $z(x, 0) \geq 0$ in $(0, 1)$ and $z(0, 0) = z(1, 0) = 0$ or the trajectory $z(x, c)$ tends to a critical point on $z = 0$ as c tends to $c_0 > 0$, which must be a saddle by the conditions on $q'(x)$. In the former case, since the direction of the trajectories across $z = 0$ is given by $-q$, we would have a root with non-negative derivative between 0 and 1, contradicting the hypothesis. In the latter case, the boundary of the band of solutions $z(x, c_0)$ must comprise the saddle and two of its trajectories as stated above. Let $(p, 0)$ denote the position of this saddle.

Now suppose for some value k in $(0, 1)$ we have $z(k, c) = c$ for all c sufficiently large. Clearly this cannot be at $k = p$ since the boundary must touch the z axis at $x = p$, but crosses the line $x = 0$ at $c_0 > 0$. Thus both of the regions $[0, k]$ and $[k, 1]$ satisfy the hypothesis, but they cannot both contain saddles. Thus for any $k \in (0, 1)$ the system cannot have a center between 0 and k or k and 1.

We now consider the return map near the trajectory $z(x, c_0)$. As the return map is analytic, the saddle can contribute no non-analytic terms, and hence it is integrable. We can now adapt the arguments of Chapter 4 to show that P and Q must be polynomials of a polynomial $A = (x - p)^2 + O((x - p)^3)$. Once again from the conditions on q, A' has only one root at $x = p$. Thus the transformation $u = \sqrt{A}\,\mathrm{sgn}(x - p)$ is well defined and analytic in $[0, 1]$, and P and Q are now polynomials of u^2.

Let $P = f(u^2)$ and $Q = g(u^2)$; then the transformation brings (6.1) to the form

$$\frac{dy}{du} = 2uf'(u^2)y^2 + 2ug'(u^2)y^3. \tag{10.3}$$

Clearly this equation is symmetric with respect to the transformation $u \to -u$. Therefore, there is a center between any two points $u = -\ell$ and $u = \ell$. Under this transformation the points $x = 0$ and $x = 1$ correspond to the values u_0 and u_1 with $u_0 < 0 < u_1$. If $-u_0 < u_1$, then it is clear that $u(k) = -u_0$ for some k in $(p, 1)$. But then we have a center of (6.1) with $a = k$ corresponding to the center of (10.3) between $-u_0$ and $u(k)$, which cannot happen; thus $-u_0 \geq u_1$. In a similar manner $-u_0 \leq u_1$ and so $-u_0 = u_1$. Hence, $A(0) = u_0^2 = u_1^2 = A(1)$ and we have proved the existence of a symmetry (the composition condition of Chapter 6) in this case. □

When the center bounds on a separatrix cycle with two saddles, then it can be shown that the two saddles have equivalent holonomies. That is, there is a map taking the neighborhood of one critical point into the other. If the holonomies are both linearizable, then we can find a local Darboux first integral. If they are not linearizable, then the maps taking one saddle to the other are more restricted and we might hope that they could be extended to a symmetry of the separatrix cycle.

Problem 10.2. Suppose a center is bounded by a separatrix cycle consisting of two curves $\ell_1 = 0$ and $\ell_2 = 0$ which intersect in two saddles p_1 and p_2. Does there exists a neighborhood U of the separatrix cycle with either a local Darboux first integral $\ell_1^\lambda \ell_2$, or a symmetry of the system $f : U \to U$ swapping p_1 and p_2?

If true, this conjecture is the simplest case of a "center-focus problem" for period annuli. If true, it would be a satisfying first step in trying to see how to extract the right sort of global information from the existence of a center.

10.3 An Experimental Approach

The final approach we want to discuss is based around some recent work of H.C. Graf v. Bothmer [19]. It allows one to obtain information about the dimensions of specific components of the center variety which are inaccessible with more conventional computer algebra.

We calculate the Lyapunov quantities as before, but now modulo a prime number p. It turns out [19] that an algorithm for computing the Lyapunov quantities can be found which works modulo p up to $L((p-3)/2)$ (after which the denominators may vanish).

Let Σ be the center variety as before (considered as an affine space again), and let $\Sigma(\mathbb{Z}_p)$ denote the points in Σ with coefficients in $\mathbb{Z}_p$. Let n_p denote the fraction of $\mathbb{Z}_p$ points in Σ compared with $\mathbb{Z}_p^N$, where N is the number of parameters. That is,

$$n_p = |\Sigma(\mathbb{Z}_p)|/p^N.$$

One consequence of the Weil conjectures [6, 74] is that

$$n_p = r\left(\frac{1}{p}\right)^c + O\left(\left(\frac{1}{p}\right)^{c+1}\right), \tag{10.4}$$

where c is the lowest codimension amongst the components of Σ and r is the number of components of that codimension.

This number can be estimated by computing a number of random points in $\mathbb{Z}_p^N$ and calculating the empirical fraction of these which lie in $\Sigma(\mathbb{Z}_p)$. We denote this number by $\tilde{n}_p$. This quantity $\tilde{n}_p$ allows us to make estimates of r and c via (10.4).

However, we would like also to be able to say something about the components of the center variety with higher codimensions. This can be done as follows. Suppose we want to calculate the number of components r' of Σ which have codimension c'. We need to exclude all points on components of Σ which have codimension less than c'. However this can be done in a nice way: given a point on $\Sigma(\mathbb{Z}_p)$ we can calculate the tangent space to $\Sigma(\mathbb{Z}_p)$ at that point. If the codimension of this tangent space is less than c' we reject this point.

The reason this works is that the codimension of the tangent space at a point on a component of Σ is always less than or equal to the codimension of the component itself. For a generic point however the codimension will be the same. Thus if we reject all points whose tangent spaces have codimension less than c', we are rejecting precisely all points on components with codimension less than c' and some non-generic points of the other components.

We calculate the fraction of points satisfying the codimension criteria above and estimate r' and c' from (10.4) as before.

This method has been shown to give accurate results for quadratic systems (where the complete classification is known) and also for cubic systems up to codimension 7, tying in with known irreducible components of the center variety [97, 132].

Higher codimension cases will require much more computational power, but the rate of growth of the complexity is much more favorable than using more standard symbolic routines.

Notes

§1 It is clear that the general solution to the center-focus problem seems very far away, in spite of a growing number of techniques and known cases.

A more realistic goal over the next few years would be to classify all centers in cubic systems. All three of the approaches suggested above could be used in this task, but some new ideas will probably be needed to bridge the still wide computational gap between what has been achieved to date, and the full complexity of cubic systems.

Perhaps this is a case where a coordinated effort by the many researchers who have an interest in this area could yield some very tangible results in the not-too-distant future.

Problem 10.3. Unite the known results on centers in cubic systems. As far as is possible, try to verify the completeness of these conditions.

A full compututional verification of the cubic centers seems unlikely with current computational power. However, a partial verfication might use finite field calculations, or restrictions of cubic systems to randomly chosen subspaces. Using the techniques of Chapter 1, it should be possible to verify that the center components are isolated and Torregrosa has been able to verify many of the center components found by Żołądek.

§2 The role of finite characteristic methods could probably be exploited more, although there are problems with just carrying across definitions from the characteristic 0 case [104].

One approach which has been used increasingly over the past years is to perform the Gröbner basis calculations for center conditions modulo a prime number. These calculations can often be performed even when the calculation over $\mathbb{Q}$ is intractable. In favourable cases, these conditions can be lifted to the original system by seeking the simplest rationals which reduce to the values of the coefficients when calculated modulo the prime. The process is described in [111].

It is also possible to calculate the Lyapunov quantities themselves using modular arithmetic as explained in Section 10.3. The size of the prime used will depend on the order of fine focus required since the expression for $L(k)$ over $\mathbb{Q}$ has a factor in the denominator of the form $(2k + 1)!!$. An interesting example of this is given in [20].

In general, it is not easy to be sure to what extent a Gröbner basis calculated modulo a prime reflects the original basis. For so-called "lucky primes" the situation is better understood, but these cannot be known beforehand. Further details can be found in [4].

Problem 10.4. With the growing ability of computers to parallelize computations, it might be interesting to implement algorithms which can calculate Gröbner basis with respect to several primes simultaneously, removing those which appear to be unlucky. Maybe such computations would be suitable for GPU processors, since the modular coefficients would not suffer from the expression swell of arbitrary length integer calculations.

Gröbner basis calculations are also being continually improved. For example, the F4 and F5 algorithms of Faugerre [62, 63].

With the growing complexity of bifurcation arguments, it is good to be able to have independent verification of the cyclicity of a particular system. Exact numerical methods can provide a computational tool which can assist in this, see [68, 79]. Methods from numerical algebraic geometry can also be used to verify center conditions [90].

It is often more expedient in the largest calculations to use a direct calculation of the factorized resultant. Multi-dimensional versions of the resultant are known but, as far as I know, have not been used in the solution of the center-focus problem and would be interesting to investigate further.

§3 The first part of Hilbert's 16th problem asks for information about the arrangement of the ovals of an algebraic curve. The second part, on limit cycles, seems to have been suggested by Hilbert as another problem where similar techniques (in fact, bifurcation methods) could be used.

Generally these two areas have proceeded quite distinctly with little overlap. However, an important modern tool in the first part of Hilbert's problem, the method of patching, can also be applied to vector fields [78]. This allows a new vector field to be constructed from several vector fields sharing the properties of both. The initial work with some non-trivial examples are given in [78], but it would be interesting to find further areas where this powerful technique can be applied.

Bibliography

[1] A. Álvarez, J-L. Bravo and C. Christopher, *On the trigonometric moment problem*, Ergodic Theory Dynam. Systems **34**, no. 1, (2014), 1–20.

[2] A. Álvarez, J-L. Bravo, C. Christopher and P. Mardešić, *Infinitesimal Center Problem on Zero Cycles and the Composition Conjecture* Func. Anal. Appl. **55**, no. 4, (2021), 3–21.

[3] A. Álvarez, J-L. Bravo and P. Mardešić, *Vanishing Abelian integrals on zero-dimensional cycles*, Proc. London Math. Soc. **107**, no. 3, (2013), 1302–1330.

[4] E.A. Arnold, *Modular algorithms for computing Gröbner bases*, J. Symb. Comput. **35**, (2003), 403–419.

[5] V. I. Arnold and Y. S. Ilyashenko, *Ordinary Differential Equations*, in *Ordinary Differential Equations and Smooth Dynamical Systems*, Spinger (1997), 1–148.

[6] J. Avellar, L.G.S. Duarte, S.E.S. Duarte and L.A.C.P. da Mota, *Integrating first-order differential equations with Liouvillian solutions via quadratures: a semi-algorithmic method*, J. Comput. Appl. Math. **182** (2005), no. 2, 327–332.

[7] J. Avellar, L.G.S. Duarte, S.E.S. Duarte and L.A.C.P. da Mota, *Finding elementary first integrals for rational second order ordinary differential equations*, J. Math. Phys. **50**, no. 1, (2009).

[8] W.A.H. Aziz, *Integrability and linearizability problems of three dimensional Lotka–Volterra equations of rank-2* Qual. Theory Dyn. Syst. **18**, (2019), 1113–1134.

[9] W.A.H Aziz and C. Christopher, *Local integrability and linearizability of three-dimensional Lotka–Volterra systems*, Appl. Math. Comput. **219**, no. 8, (2012), 4067–4081.

[10] W.A.H Aziz and C. Christopher, *On the integrability of some three-dimensional Lotka-Volterra equations with rank-1 resonances* Publ. Mat. **58**, no. 1, (2014), 37–48.

[11] W.A.H Aziz, C. Christopher, J. Llibre and C. Pantazi, *Three-Dimensional Lotka–Volterra Systems with 3:-1:2 Resonance*. Mediterr. J. Math. **18**, 167, (2021).

© The Author(s), under exclusive license to Springer Nature Switzerland AG 2024
C. Christopher et al., *Limit Cycles of Differential Equations*, Advanced Courses in Mathematics - CRM Barcelona, https://doi.org/10.1007/978-3-030-59656-9

[12] W.A.H Aziz, C. Christopher and R. Oliveira, *The monodromy method and integrability of three dimensional vector fields of Lotka–Volterra type*, In preparation.

[13] W.A.H. Aziz, C. Christopher, C. Pantazi and S. Walcher, *Liouvillian integrability of three dimensional vector fields*, in preparation.

[14] N. N. Bautin, *On the number of limit cycles which appear with the variation of coefficients from an equilibrium position of focus or center type.* Amer. Math. Soc. Translation **100** (1954), 19 pp.

[15] L.R. Berrone and H. Giacomini, *Inverse Jacobi multipliers*, Rend. Circ. Mat. Palermo **52**, (2003), 77–130.

[16] M. Berthier, D. Cerveau and A. Lins Neto, *Sur les feuilletages analytiques réels et le problème du centre*, J. Differential Equations **131** (1996), no. 2, 244–266.

[17] M. Bobieński and H. Żołądek, *The three-dimensional generalized Lotka-Volterra systems*, Ergod. Theory. Dynam. Syst. **25**, (2005), 759–791.

[18] P. Bonnet, *Description of the module of relatively exact 1-forms modulo a polynomial f on $\mathbb{C}^2$*, Preprint, Université de Bourgogne (1999), no.184.

[19] H.C.G. von Bothmer, *Experimental results for the Poincaré center problem*, Nonlinear differ. equ. appl. **14**, (2007), 671–698.

[20] H.C.G. von Bothmer and Jakob Kröker, *Focal values of plane cubic centers*, Qual. Theory Dyn. Syst. **9**, (2010), 319–324.

[21] M. Briskin, J.-P. Francoise and Y. Yomdin, *Center conditions, compositions of polynomials and moments on algebraic curves*, Ergodic Theory Dynam. Systems **19**, no. 5, (1999), 1201–1220.

[22] M. Briskin, J.-P. Francoise and Y. Yomdin, *Center conditions. II. Parametric and model center problems*, Israel J. Math. **118** (2000), 61–82.

[23] M. Briskin, J.-P. Francoise and Y. Yomdin, *Center conditions. III. Parametric and model center problems*, Israel J. Math. **118** (2000), 83–108.

[24] M. Briskin, N. Roytvarf and Y. Yomdin, *Center conditions at infinity for Abel differential equations*, Ann. of Math. **172**, no. 2, (2010), 437–483.

[25] M. Briskin, F. Pakovich and Y. Yomdin, *Algebraic geometry of the center-focus problem for Abel differential equations*, Ergodic Theory Dynam. Systems **36**, no. 3, (2016), 714–744.

[26] C. Bujac and N. Vulpe, *Cubic differential systems with invariant straight lines of total multiplicity eight and four distinct infinite singularities*, J. Math. Anal. **423**, no. 2, (2015), 1025–1080.

[27] L. Cairó, H. Giacomini and J. Llibre, *Liouvillian first integrals for the planar Lotka-Volterra system*, Rend. Circ. Mat. Palermo **52**, no. 3, (2003), 389–418.

[28] M. Carnicer, *The Poincaré Problem in the Nondicritical Case*, Ann. Math. **140**, no. 2, (1994), 289–294.

[29] G. Casale, *Suites de Godbillon-Vey et intégrales premières*, C. R. Math. Acad. Sci. Paris **335** (2002), no. 12, 1003–1006.

[30] D. Cerveau and A. Lins Neto, *Holomorphic foliations in* $\mathbb{CP}(2)$ *having an invariant algebraic curve*, Ann. Inst. Fourier **41**, no. 4, (1991), 883–903.

[31] D. Cerveau and A. Lins Neto, *Irreducible components of the space of holomorphic foliations of degree two in* $\mathbb{CP}(n)$, $n \geq 3$, Ann. of Math. **143** (1996), no. 3, 577–612.

[32] D. Cerveau, A. Lins-Neto, F. Loray, J.V. Pereira and F. Touzet, *Algebraic Reduction Theorem for complex codimension one singular foliations*. Comment. Math. Helv. **81**, no. 1, (2006), 157–169.

[33] D. Cerveau, A. Lins-Neto, F. Loray, J.V. Pereira and F. Touzet, *Complex codimension one singular foliations and Godbillon–Vey sequences*, Mosc. Math. J. **7**, no. 1, (2007), 21–54.

[34] J. Chavarriga, J. LLibre and J. Sotomayor, *Algebraic solutions for polynomial systems with emphasis in the quadratic case*, Expositiones Mathematicae **15** (1997), 161–173.

[35] L.A. Cherkas, *On the conditions for a centre for certain equations of the form* $yy' = P(x) + Q(x)y + R(x)y^2$, Differential Equations **8**, no. 8, (1974), 1104–1107.

[36] C. Chicone and M. Jacobs, *Bifurcation of critical periods for plane vector fields*, Trans. Amer. Math. Soc. **312**, (1989), 433–486.

[37] C. Christopher, *An algebraic approach to the classification of centers in polynomial Liénard systems*, J. Math. Anal. Appl. **229**, no. 1, (1999), 319–329.

[38] C. Christopher, *Abel equations: composition conjectures and the model problem*, Bull. London Math. Soc. **32**, no. 3, (2000), 332–338.

[39] C. Christopher, *Estimating limit cycle bifurcations from centers*, in *Differential Equations with Symbolic Computation* (ed. D. Wang and Z. Zheng), Springer (2005), 23–36.

[40] C. Christopher and J. Devlin, *On the classification of Liénard systems with amplitude-independent periods*, J. Differ. Equ. **200**, no. 1, (2004), 1–17.

[41] C. Christopher, J. Giné, Y. Tang and J. Torregrosa, *Local integrability of generalized Kukles systems*, In preparation.

[42] C. Christopher, W.M.A. Hussein and Z. Wang, *On the Integrability of Lotka–Volterra Equations: An Update*, in *Mathematical Sciences with Multidisciplinary Applications* (ed. B. Toni), Springer (2016).

[43] C. Christopher and J. Llibre, *Algebraic aspects of Integrability for polynomial systems*, Qualitative Theory of Dynamical Systems **1** (1999), 71–95.

[44] C. Christopher, and J. Llibre, *A family of quadratic differential systems with invariant algebraic curves of arbitrarily high degree without rational first integrals*, Proc. Am. Math. Soc. **7**, (2001), 2025–2030.

[45] C. Christopher, J. Llibre, C. Pantazi and X. Ziang, *Darboux integrability and invariant algebraic curves for planar polynomial systems*, J. Physics A: Gen. Math. **35** (2002), 2457–2476.

[46] C. Christopher, J. Llibre, C. Pantazi and S. Walcher, *Inverse problems for multiple invariant curves.* Proc. Roy. Soc. Edinburgh A **137**, no. 6, (2007), 1197–1226.

[47] C. Christopher, J. Llibre, C. Pantazi and S. Walcher, *Inverse problems for invariant algebraic curves: Explicit computations.* Proc. Roy. Soc. Edinburgh A **139** no. 2, (2009), 287–302.

[48] C. Christopher, J. Llibre, C. Pantazi and S. Walcher, *Darboux integrating factors: inverse problems*, J. Differ. Equ. **250**, no. 1, (2011), 1–25.

[49] C. Christopher, J. Llibre, C. Pantazi and S. Walcher, *Inverse problems in Darboux' theory of integration*, Acta Appl. Math. **120**, (2012), 101–126.

[50] C. Christopher, J. Llibre, C. Pantazi and S. Walcher, *On planar polynomial vector fields with elementary first integrals*, J. Differ. Equ. **267**, no. 8, (2019), 4572–4588.

[51] C. Christopher, J. Llibre and J.V. Pereira, *Multiplicity of invariant algebraic curves in polynomial vector fields*, Pacific J. Math. **229**, no. 1, (2007), 63-117.

[52] C. Christopher, J. Llibre and G. Świrszcz, *Invariant algebraic curves of large degree for quadratic system*, J. Math. Anal. Appl. **303**, no. 2, (2005), 450–461.

[53] C. Christopher and S. Lynch, *Small-amplitude limit cycle bifurcations for Liénard systems with quadratic or cubic damping or restoring forces*, Nonlinearity **12**, no. 4, (1999), 1099–1112.

[54] C. Christopher and P. Mardešić, *The monodromy problem and the tangential center Problem*, Func. Anal. Appl. **44**, no. 1, (2010), 22–35.

[55] C. Christopher and P. Mardešić, *Darboux relative exactness and pseudo-Abelian integrals*, in preparation.

[56] C. Christopher, P. Mardešić, and C. Rousseau, *Normalizable, integrable, and linearizable saddle points for complex quadratic systems in $\mathbb{C}^2$* J. Dyn. Control Syst. **9**, no. 3 (2003), 311–363.

[57] C. Christopher and C. Rousseau, *Normalizable, integrable and linearizable saddle points in the Lotka-Volterra system*, Qualit. Theory of Dynam. Syst. **5** (2004), 11–61.

[58] C. Christopher and D. Scholmiuk, *Center conditions for a class of polynomial differential systems*, CRM Preprint series, CRM-3236, Université de Montréal (2006).

[59] P. Deligne, *La conjecture de Weil. I.*, Inst. Hautes Études Sci. Publ. Math. **43** (1974), 273–307.

[60] J. D. Dixon and B. Mortimer, *Permutation groups*, Graduate Texts in Mathematics **163**, Springer-Verlag, New York (1996).

[61] S. Evdokimov and I. Ponomarenko, *A new look at the Burnside-Schur theorem*, Bull. London Math. Soc. **37** (2005), 535–546.

[62] J.-C. Faugère, *A new efficient algorithm for computing Gröbner bases (F4)* J. Pure and Appl. Algebra **139**, no. 1, (1999), 61–88.

[63] J.-C. Faugère, *A new efficient algorithm for computing Gröbner bases without reduction to zero (F5)* in *Proceedings of the 2002 International Symposium on Symbolic and Algebraic Computation (ISSAC)*, ACM Press, (2002), 75–83.

[64] O. Forster, *Lectures on Riemann surfaces*, Graduate Texts in Mathematics **81**, Springer-Verlag, New York-Berlin (1981).

[65] J.-P. Françoise, *Successive derivatives of a first return map, application to the study of quadratic vector fields*, Ergodic Theory Dynam. Systems **16**, no. 1, (1996), 87–96.

[66] J.-P. Françoise, N. Roytvarf and Y. Yomdin, *Analytic continutation and fixed points of the Poincaré mapping for a polynomial Abel equation*, J. Eur. Math. Soc. **10**, no. 2, (2008), 543–570.

[67] A. Fronville, A. Sadovski and H. Żołądek, *Solution of the 1:2 resonant center problem in the quadratic case*, Fundamenta Mathematicae **157** (1998), 191–207.

[68] Z. Galias and W. Tucker, *The Songling system has exactly four limit cycles*, Appl. Math. Comput. **415**, (2022), 126691.

[69] L. Gavrilov and H. Movasati, *The infinitesimal 16th Hilbert problem in dimension zero*, Bull. Sci. Math. **131**, no. 3, (2007), 242–257.

[70] J. Giné and M. Grau, *Characterization of isochronous foci for planar analytic differential systems*, Proc. Roy. Soc. Edinburgh Sect A **135**, no. 5, (2005), 985–998.

[71] J. Giné, M. Grau and X. Santalluisa, *A counterexample to the composition condition conjecture for polynomial Abel differential equations.* Ergod. Theory Dyn. Syst. **39**, no. 12, (2019), 3347–3352.

[72] J. Giné and J. Llibre, *On the integrability of Liénard systems with a strong saddle*, Appl. Math. Lett. **70**, (2017), 39–45.

[73] S. Gravel and P. Thibault, *Integrability and linearizability of the Lotka-Volterra system for $\lambda \in \mathbb{Q}$*, J. Differential Equations **184** (2002), 20–47.

[74] R. Hartshorne, *Algebraic geometry*, Graduate Texts in Mathematics **52**, Springer-Verlag, New York-Heidelberg (1977).

[75] W.M.A. Hussein and C. Christopher, *A Geometric Investigation of the Invariant Algebraic Curves in Two Dimensional Lotka–Volterra Systems*, Math. Comput. Sci. **11**, (2017), 269–283.

[76] Y. Il'yashenko, *The origin of limit cycles under perturbation of the equation $dw/dz = -R_z/R_w$, where $R(z, w)$ is a polynomial*, Mat. USSR Sbornik **7**, no. 3, (1969), 353–364.

[77] Y. Il'yashenko and S. Yakovenko, *Lectures on Analytic Differntial Equations*, American Mathematical Society, (2009)

[78] I. Itenberg and E. Shustin, *Singular points and limit cycles of planar polynomial vector fields*, Duke Math. J. **102**, no. 1, (2000), 1–37.

[79] T. Johnson and W. Tucker, *A rigorous study of possible configurations of limit cycles bifurcating from a hyper-elliptic Hamiltonian of degree five*, Dyn. Syst. **24**, no. 2, (2009), 237–247.

[80] P. Joyal, *Generalised hopf bifurcations and its dual generalised homoclinic bifurcation*, SIAM J. Appl. Math. **48**, no.3, (1988), 481–496.

[81] P. Joyal and C. Rousseau, *Saddle quantities and applications*, J. Differ. Equ., **78**, no. 2, (1989),374–399.

[82] H. Liang and J. Torregrosa, *Parallelization of the Lyapunov constants and cyclicity for centers of planar polynomial vector fields*, J. Differ. Equ. **259**, no. 11, (2015), 6494–6509.

[83] H. Liang and J. Torregrosa, *Weak-Foci of High Order and Cyclicity*, Qual. Theory Dyn. Syst. **16**, (2017), 235–248.

[84] A. Lins Neto, *Algebraic solutions of polynomial differential equations and foliations in dimension two*, in Holomorphic Dynamics, Lect. Notes in Math. **135** (1988), 192–232.

[85] C. Liu, G. Chen and C. Li, *Integrability and linearizability of the Lotka-Volterra systems*, J. Differ. Equ. **198**, no. 2, (2004), 301–320.

[86] J. Llibre and X. Zhang, *Darboux theory of integrability in $\mathbb{C}^n$ taking into account the multiplicity*, J. Differ. Equ. **246**, no. 2, (2009), 541–551.

[87] D. López García, *The monodromy problem for hyperelliptic curves*, Bull. Sci. Math. **170**, (2021), 102998.

[88] D. López García, *Monodromy problem and tangential center-focus problem for product of generic lines in* $\mathbb{P}^2$, Preprint.
https://arxiv.org/pdf/2205.03496.pdf

[89] F. Loray, *A preparation theorem for codimension-1 foliations*, Ann. Math. **163**, (2006), 709–722.

[90] A. Mahdi, C. Pessoa and J.D. Hauenstein, *A hybrid symbolic-numerical approach to the center-focus problem*, J. Symb. Comput. **82**, (2017), 57–73.

[91] Y-K. Man and M. A. H. MacCallum, *A rational approach to the Prelle-Singer algorithm*, J. Symb. Comput. **24** (1997), 31–43.

[92] P. Mardešić, D. Marín and J. Villadelprat, *The period function of reversible quadratic centers*, J. Differ. Equ., **224**, no. 1, (2006), 120–171.

[93] P. Mardešić, C. Rousseau and B. Toni, *Linearization of isochronous centers*, J. Differ. Equ. **121** (1995), 67–108.

[94] J. Martinet and J.-P. Ramis *Classification analytique des équations non linéaires résonnantes du premier ordre*, Ann. Scient. Éc. Norm. Sup. 4e série **16** (1983), 571–621.

[95] J.-F. Mattei and R. Moussu, *Holonomie et intégrales premières*, Ann. Scient. Éc. Norm. Sup. 4e série **13** (1980), 469–523.

[96] J. Moulin-Ollagnier, *Liouvillian integration of the Lotka-Volterra system*, Qualitative Theory of Dynamical Systems, **3**, (2002), 19–28.

[97] H. Movasati, *Center conditions: rigidity of logarithmic differential equations*, J. Differential Equations **197** (2004), 197–217.

[98] P. Müller, *Primitive monodromy groups of polynomials.* Recent developments in the inverse Galois problem (Seattle, WA, 1993), 385–401, Contemp. Math., **186**, Amer. Math. Soc., Providence, RI, (1995).

[99] H. M. Osinga and J. Moehlis, *A continuation method for computing global isochrons*, SIAM J. Appl. Dyn. Syst. **9** no. 4 (2010), 1201–1228.

[100] F. Pakovich, *A counterexample to the "composition conjecture"*, Proc. Amer. Math. Soc. **130**, no. 12, (2002), 3747–3749.

[101] F. Pakovich, *Solution of the parametric center problem for the Abel differential equation*, J. Eur. Math. Soc. **19**, no. 8, (2017), 2343–2369.

[102] F. Pakovich and M. Musychuk, *Solution of the polynomial moment problem*, Proc. Lond. Math. Soc. **99**, no. 3, (2009), 633–657.

[103] J. V. Pereira, *Integrabilidade de Equações Diferenciais no Plano Complexo*, Monografias del IMCA **25** (2002).

[104] J. V. Pereira, *Invariant hypersurfaces for positive characteristic vector fields*, J. Pure Appl. Algebra **171**, no. 2–3, (2002), 295–301.

[105] R. Pérez-Marco and J.-C. Yoccoz, *Germes de feuilletages holomorphes à holonomie prescrite*, Astérisque **222** (1994), 345–371.

[106] R. Prohens and J. Torregrosa, *New lower bounds for the Hilbert numbers using reversible centers* Nonlinearity **32**, (2019), 331–355.

[107] J.F. Ritt, *Prime and composite polynomials*, Trans. Amer. Math. Soc. **23**, no. 1, (1922), 51—66.

[108] M.J. Prelle and M.F. Singer, *Elementary first integrals of differential equations*, Trans. Amer. Math. Soc. **279** (1983), 619–636.

[109] V.G. Romanovski, Y.H. Wu and C. Christopher, *Linearizability of three families of cubic systems* J. Phys. A: Math. Gen. **36** (2003), L145–L152.

[110] V.G. Romanovski and D.S. Shafer, *The Centre and Cyclicity Problems: A Computational Algebra Approach*, Birkhäuser (2009).

[111] V.G. Romanovski and D.S. Shafer, *Centers and limit cycles in polynomial systems of ordinary differential equations*, in *School on real and complex singularities in São Carlos, 2012*, Adv. Stud. Pure Math. **68**, (2016), 267–373.

[112] M. Rosenlicht, *On the explicit solvability of certain transcendental equations*, Inst. Hautes Études Sci. Publ. Math. **36** (1969), 15–22.

[113] R. Roussarie, *On the number of limit cycles which appear by perturbation of separatrix loop of planar vector fields*, Bol. Soc. Brasil. Mat. **17** (1986), 67–101.

[114] R. Roussarie, *Bifurcation of planar vector fields and Hilbert's sixteenth problem*, Progress in Mathematics **164**, Birkhäuser Verlag, Basel (1998).

[115] C. Rousseau and D. Schlomiuk, *Cubic vector fields symmetric with respect to a center*, J. Differential Equations **123** (1995), 388–436.

[116] R. Salih and C. Christopher, *Center bifurcations of 3D Lotka Volterra systems*, in preparation.

[117] D. Schlomiuk, *Algebraic particular integrals, integrability and the problem of the centre*, Trans. Amer. Math. Soc. **338** (1993), 799–841.

[118] W.R. Scott, *Group Theory*, Dover (2003).

[119] M.F. Singer, *Liouvillian first integrals of differential equations*, Trans. Amer. Math. Soc. **333** (1992), 673–688.

[120] D. Schlomiuk and N. Vulpe, *Planar quadratic differential systems with invariant straight lines of total multiplicity four*, Nonlinear Anal. Theory Methods Appl. **68**, no. 4, (2008), 681–715.

[121] E. Stróżyna and H. Żołądek, *The analytic and formal normal form for the nilpotent singularity*, J. Differ. Equ. **179**, no. 2, (2002), 479–537.

[122] M. Uribe, *Principal Poincaré-Pontryagin function associated to polynomial perturbations of a product of* $(d + 1)$ *straight lines*, J. Differ. Equ. **246**, no. 4 (2009), 1313–1341.

[123] Q. Wang, Y. Liu and C. Haibo, *Hopf bifurcation for a class of three-dimensional nonlinear dynamic systems*, Bull. Sci. Math. **134**, no. 7, (2010), 786–798.

[124] Q. Wang, J. Lu, W. Huang *Integrability and linearizability for Lotka-Volterra systems with the 3:-q resonant saddle point*, Adv. Difference Equations **2014**, (2014), 1–15.

[125] Q. Wang, J. Lu, W. Huang and B. Sang, *The Center Conditions and Hopf Cyclicity for a 3D Lotka-Volterra System*, J. Nonlinear Modeling and Analysis **3**, no. 1, (2021), 1–12.

[126] H. Wielandt, *Finite permutation groups*, Academic Press (1964).

[127] Y. Zare and S. Tanabé, *Brieskorn Module and Center Conditions: Pull-Back of Differential Equations in Projective Space*, J. Dyn. Control Syst. **27**, (2021), 799–816.

[128] X. Zhang, *Integrability of Dynamical Systems: Algebra and Analysis*, Springer (2017).

[129] H. Żołądek, *On a certain generalization of Bautin's theorem*, Nonlinearity **7**, no. 1, (1994), 273–279.

[130] H. Żołądek, *The classification of reversible cubic systems with center*, Topol. Methods Nonlinear Anal. **4**, no. 1, (1994), 79–136.

[131] H. Żołądek, *Eleven small limit cycles in a cubic vector field*, Nonlinearity **8**, no. 5, (1995), 843–860.

[132] H. Żołądek, *Remarks on: "The classification of reversible cubic systems with center"*, Topol. Methods Nonlinear Anal. **8**, no. 2, (1996), 335–342.

[133] H. Żołądek, *The problem of center for resonant singular points of polynomial vector fields*, J. Differ. Equ. **137**, (1997), 94–118.

[134] H. Żołądek, *The extended monodromy group and Liouvillian first integrals*, J. Dynamical and Control Systems **4**, (1998), 1–28.

[135] H. Żołądek, *The Monodromy Group*, Birkhäuser (2006).

Part II

Abelian Integrals and Applications to the Weak Hilbert's 16th Problem

Chengzhi Li & Joan Torregrosa

Preface of the first edition

The second part of Hilbert's 16th problem, asking for the maximum $\mathcal{H}(n)$ of the numbers of limit cycles and their relative positions for all planar polynomial differential systems of degree n, is still open even for the quadratic case ($n = 2$).

A weak form of this problem, proposed by V.I. Arnold, asking for the maximum $Z(m, n)$ of the numbers of isolated zeros of Abelian integrals of all polynomial 1-forms of degree n over algebraic ovals of degree m, is also extremely hard to grasp. The number $\tilde{Z}(n) = Z(n+1, n)$ can be chosen as a lower bound of $\mathcal{H}(n)$; so far only $\tilde{Z}(2) = 2$ has been proved.

These lecture notes are devoted to the introduction of some basic concepts and methods in the study of Abelian integrals and applications to the weak Hilbert's 16th problem. In Chapter 1 we briefly introduce Hilbert's 16th problem and its weak form. In Chapter 2 we explain the relation between the study of Abelian integrals and the study of limit cycles. In Chapter 3 we use several methods to study the number of zeros of the Abelian integrals associated with perturbations of the Bogdanov–Takens system. At last, in Chapter 4 we introduce a proof of $\tilde{Z}(2) = 2$, the method of the proof is unified for all regions of the parameter space.

I would like to express my sincere appreciation to Jaume Llibre and Armengol Gasull for their kind invitation and collaborations, to Armengol Gasull, Iliya D. Iliev, Weigu Li, Jaume Llibre, Dana Schlomiuk, Zhifen Zhang, and Yulin Zhao, who carefully read the manuscript of the notes, and provided valuable comments and corrections. I also want to thank the director, Manuel Castellet, and all staff of CRM for their excellent assistance and support. I am grateful to my colleagues at Peking University for numerous discussions and cooperation, especially to Zhifen Zhang and Tongren Ding who led me to the research field many years ago and never ceased to encourage me.

Chengzhi Li

Preface of the second edition

The main focus and structure of the book have been preserved in this second edition. I am grateful for the positive reception of the first edition and constructive comments by friends and colleagues. During the almost 14 years since the first edition, I have received numerous email messages from readers with comments on the lectures and some suggestions and new references that would be added if a new edition were proposed. With the aid of all this information, together with Joan Torregrosa, we have completely revised the notes on this advanced course on Abelian Integrals.

Chengzhi Li

This edition provides technical corrections, misprints, updates, and a not short list of clarifications and remarks in the four chapters, but mainly in the first three. The aim has been to clarify the results explained during the lessons of the advanced course that took place in June of 2006 in Barcelona. Adding, in particular, a short review of the new results on the lower bounds of $\mathcal{H}(n)$ and some applications of the averaging bifurcation technique. The most obvious change is the big list of references that have been doubled since the first edition.

We would like to express our many thanks to Professors Jaume Llibre, Armengol Gasull, and Iliya D. Iliev, as well as our Chinese colleagues Hebai Chen, Maoan Han, Xiaochun Hong, Wentao Huang, Feng Li, Jibin Li, Haihua Liang, Changjian Liu, Yirong Liu, Linping Peng, Yilei Tang, Yonghui Xia, Dongmei Xiao, Jiang Yu, Pei Yu, Xiang Zhang, Liqin Zhao, and Yulin Zhao for their valuable comments and kindly providing references.

Chengzhi Li & Joan Torregrosa

Chapter 1

Hilbert's 16th Problem and Its Weak Form

1.1 Hilbert's 16th Problem

Consider the planar differential system

$$\dot{x} = P_n(x, y), \quad \dot{y} = Q_n(x, y), \tag{1.1}$$

where P_n and Q_n are real polynomials in x, y with maximum degree n. The second half of the famous Hilbert's 16th problem, proposed in 1900, can be stated as follows (see [147]):

For a given integer n, what is the maximum number of limit cycles of system (1.1) for all possible P_n and Q_n? And how about the possible relative positions of the limit cycles?

Usually, the maximum of the number of limit cycles is denoted by $\mathcal{H}(n)$, and is called the Hilbert number. We recall that a *limit cycle* of system (1.1) is an isolated closed orbit. It is the α-(backward) or ω-(forward) limit set of nearby orbits. In many applications, the number and positions of them are important to understand the global dynamics of the corresponding system. Note that the problem is trivial for $n = 1$, because a linear system may have periodic orbits but no limit cycles, so we assume $n \geq 2$.

This problem is still open even for the case $n = 2$, and there is no answer even to the question of whether $\mathcal{H}(2)$ is finite or not. In [312] S. Smale said: "Except for the Riemann hypothesis it seems to be the most elusive of Hilbert's problems" (see also [311]).

Below we detail some results, among a long list of works on this problem.

© The Author(s), under exclusive license to Springer Nature Switzerland AG 2024
C. Christopher et al., *Limit Cycles of Differential Equations*, Advanced Courses
in Mathematics - CRM Barcelona, https://doi.org/10.1007/978-3-030-59656-9_11

1.1.1   The finiteness problem

- In 1923 H. Dulac [74] claimed the *individual finiteness* of limit cycles, i.e. for a *given* system (1.1) the number of limit cycles is finite. A gap in his arguments was found in the early 1980s.

- In 1985 R. Bamón [10] proved this individual finiteness property for the quadratic case $(n = 2)$.

- In the early 1990s J. Écalle and Y. Ilyashenko published, independently in two long papers [88] and [166], new proofs of the individual finiteness theorem, filling up the gap in Dulac's paper. This "is the most spectacular and the most general fact established so far in connection with the Hilbert 16th problem" (see S. Yakovenko [345]); and "these two papers have yet to be thoroughly digested by the mathematical community" (see S. Smale [312]). Naturally, the next step is to prove the *uniform* finiteness, i.e. $\mathcal{H}(n) < \infty$. Recently, Y. Ilyashenko has found some gaps in his initial proof, see [168].

- In 1988 R. Roussarie [292] proposed a program for proving the uniform finiteness by reducing this problem, via the compactification of the systems and of the parameter space, to prove the finite cyclicity of limit periodic sets (see also the work of J.-P. Françoise & C.C. Pugh [91]). F. Dumortier, R. Roussarie & C. Rousseau in [85, 86] started the program for the quadratic case and listed 121 graphics as all limit periodic sets which are necessary in this study. A series of papers, among them [75, 76, 77, 155, 295, 296, 391], continues this program, and about 85 of the 121 graphics have been studied. The remaining graphics are more degenerate and their study is more difficult.

    For a detailed introduction to the finiteness problem we refer to an article by D. Schlomiuk (the first chapter of [303]).

1.1.2   Configuration of limit cycles

There are many papers dealing with the relative positions of limit cycles for system (1.1). General results were obtained by J. Llibre & G. Rodríguez [241] in 2004 and by B. Coll, F. Dumortier & R. Prohens [62] in 2013. Let us briefly introduce such results.

A *configuration of limit cycles* is a finite set $C = \{C_1, \ldots, C_m\}$ of disjoint simple closed curves of the plane such that $C_i \cap C_j = \emptyset$ for all $i \neq j$. Given a configuration of limit cycles $C = \{C_1, \ldots, C_m\}$ the curve C_i is *primary* if there is no curve C_j of C contained in the bounded region limited by C_i. Two configurations of limit cycles C and $\hat{C}$ are (topologically) *equivalent* if there is a homeomorphism in $\mathbb{R}^2$, mapping C to $\hat{C}$.

A system (1.1) *realizes* the configuration of limit cycles C if the set of all its limit cycles is equivalent to C.

Theorem 1.1 ([241]). *Let C be a configuration of limit cycles, and let r be its number of primary curves. Then $C = \{C_1, \ldots, C_m\}$ is realizable as algebraic limit cycles by a polynomial system (1.1) of degree $n \leq 2(m+r) - 1$.*

Theorem 1.2 ([62]). *For any given configuration of n disjoint simple closed curves in the plane there exists a polynomial $F(x)$ of degree $2n+1$ and a polynomial $g(x)$ of degree $2r - 1$, for some $r \leq n$, such that the polynomial Liénard equation*

$$\dot{x} = y - F(x),$$
$$\dot{y} = \varepsilon g(x),$$

has, for $\varepsilon \gtrsim 0$, a phase portrait with $r - 1$ saddles, r foci, and a configuration of limit cycles that is topologically equivalent to the given configuration of simple closed curves.

The vector field in Theorem 1.1 is explicitly constructed being the curves C_i given circles, in a topologically equivalent configuration of the originally prescribed one. This is why the authors say that the configuration is realizable as algebraic limit cycles. Subsequently, a new and easier construction was obtained by D. Peralta-Salas [272] in 2005. Clearly, Theorem 1.2 is a perturbative result and the fixed configuration is found near a slow-fast system in the Liénard polynomial class but the degree is decreased.

These theorems can be seen as partial answers to the position question of the second part of Hilbert's 16th problem. The remaining question is: *For a fixed integer n what kinds of configurations of limit cycles of systems (1.1) are possible?* In the next subsection there is some information about this question in the quadratic family.

1.1.3 Some results on quadratic systems

- In 1952 N.N. Bautin [11] proved a fundamental fact for quadratic systems: at most three limit cycles can bifurcate from a weak focus or a center of system (1.1) for $n = 2$. A *weak focus* means a focus at which the linear part of the system is of center type. See [89] for a computation of the focal values (also known as Lyapunov constants) in Bautin's formula.

- In 1955 I.G. Petrovskii & E.M. Landis [279] attempted to prove that "$\mathcal{H}(2) = 3$".

- In 1959 C. Tung [326] found some important properties of quadratic systems: a closed orbit is convex; there is a unique singularity in the interior of it; two closed orbits are similarly (resp. oppositely) oriented if their interiors have (resp. do not have) common points. Hence, the distributions of limit cycles of quadratic systems have only one or two nests.

- In the 1960s Y. Ye classified the quadratic systems into three classes using that anyone having limit cycle(s) can be transformed into the form $\dot{x} = -y + \delta x + lx^2 + mxy + ny^2, \dot{y} = x(1 + ax + by)$. In particular, it belongs to class I if $a = b = 0$, to class II if $b = 0, a \neq 0$, and to class III if $b \neq 0$. It was proved in [41, 353] that at most one limit cycle exists for systems in class I. See also [355].

- In 1979 S. Shi [308, 309], L. Chen & M. Wang [40] found counter-examples to the result of Petrovskii–Landis. In both examples the four limit cycles are located in two nests, with at least three in one nest and at least one in another. We say that they are in $(3, 1)$-distribution.

- In 1986 C. Li [178] proved that there is no limit cycle surrounding a weak focus of third-order for any quadratic system, which gives no possibility to construct $(4, 0)$-distribution of limit cycles by perturbing a quadratic system with a weak focus of third-order, because there is no limit cycle surrounding the focus before perturbation.

- Around 2000, in a series of papers (see [366, 367]), P. Zhang proved that there is at most one limit cycle surrounding a weak focus of second-order for any quadratic system, and if a quadratic system has two nests of limit cycles, then at least one nest contains a unique limit cycle, hence $(2, 2)$-distribution of limit cycles for a quadratic system is impossible. The uniqueness of limit cycle was proved also in [66] for quadratic systems having an invariant straight line, see also [390]. The same property holds when the quadratic system has exactly two finite real singularities, a weak focus and a strong focus, and at least two singularities at infinity, see [175]. In a recent paper, J. Llibre & X. Zhang [246] have revisited the non-existence, existence and uniqueness of limit cycles problem for quadratic systems.

- There is a series of papers towards a systematic study of the global geometry of quadratic differential systems. Among them, R. Roussarie & D. Schlomiuk [294] and D. Schlomiuk [302] give a general framework of study of the class of all quadratic systems; D. Schlomiuk & N. Vulpe [304] study the geometry of quadratic systems in the neighborhood of infinity; also in the paper [302] D. Schlomiuk gives a short history of invariants theory and motivation for using them in the global theory. J. Llibre & D. Schlomiuk [242] determines the global geometry of quadratic systems with a weak focus of third-order. J.C. Artés, J. Llibre & D. Schlomiuk [8] make a global study of the closure of systems having a weak focus of second-order within quadratic systems. The global geometry of this class reveals interesting bifurcation phenomena; for example, all phase portraits with limit cycles obtained in this class can be produced by perturbations of symmetric (reversible) quadratic systems with a center. The global topological configurations of singularities for the whole family of quadratic differential systems is finished recently by J.C. Artés, J. Llibre, D. Schlomiuk & N. Vulpe [9].

Today most mathematicians in this field believe that $\mathcal{H}(2) = 4$. But in addition the possible configurations are also an open problem because nobody has shown that the configuration $(4, 0)$ is impossible to find. Only that it is far from a system having a singularity of third-order weak focus type. The last item explains the importance of the study of perturbations of centers in the weak Hilbert's 16th problem, which is the main topic of this monograph.

1.1.4 Some results on cubic and higher degree systems

- In 1954 K.S. Sibirskiĭ [310] studied the Hopf bifurcation for cubic systems without quadratic terms. This class of systems was studied again by T.R. Blows & N. Lloyd [14] 30 years later, but the complete proof that at most five limit cycles can appear was obtained by H. Żołądek [392] in 1994. In 1987 J. Li & Q. Huang [205] constructed an example showing that $\mathcal{H}(3) \geq 11$. The 11 limit cycles form "compound eyes": a big limit cycle surrounds two smaller limit cycles, each of them surrounds two nests, with at least two limit cycles in each nest. J. Li & Z. Liu [209] and M. Han, T. Zhang & H. Zang [142] also gave examples to show that $\mathcal{H}(3) \geq 11$. In 2005 P. Yu & M. Han [357] and Y. Liu & W. Huang [228] gave examples of $\mathcal{H}(3) \geq 12$, both with a $(6, 6)$-distribution of limit cycles. C. Li, C. Liu & J. Yang [187] and Y. Liu & J. Li [234] gave examples of $\mathcal{H}(3) \geq 13$, with different configurations and surrounding different number of singularities.

- In 1995 H. Żołądek [394] proved that surrounding a singularity of a cubic system there may exist 11 limit cycles by using the second-order Melnikov function. In 2016 Y. Tian & P. Yu [324] showed that, up to first-order Taylor developments, the basis chosen in the proof by Żołądek were not independent, which leads to failure in drawing the conclusion of the existence of 11 limit cycles; in fact only 9 limit cycles can exist, agreeing with the result by P. Yu & M. Han [358]. A recent paper by Y. Tian & P. Yu [325] shows that the system studied by H. Żołądek can have 10 limit cycles under perturbations up to third-order of a small parameter ε and 11 under perturbations at least up to seventh-order. All these limit cycles are of small amplitude and bifurcate from the center itself. This result has been confirmed by L.F. Gouveia & J. Torregrosa in [127]. On the other hand, C. Christopher [56] in 2005 and Y. Bondar' & A. Sadovskiĭ [16] in 2008 gave two examples of cubic systems with 11 small limit cycles surrounding a singularity, they bifurcate from a center having a rational first integral. Recently, also in [127], it is shown that these number of limit cycles appear also perturbing other cubic centers of Darboux type which are listed in [395]. In 2014 P. Yu & Y. Tian [362] gave an example with 12 limit cycles surrounding a singularity for cubic systems, they appear perturbing a cubic system of Darboux type. The proof had some gaps that were corrected by J. Giné, L.F. Gouveia & J. Torregrosa in [121].

- In the above two items we have described the known results about the maximal number $M(n)$ of small amplitude limit cycles bifurcating from an elementary center or an elementary focus, for polynomial systems of degrees 2 and 3. Providing $M(2) = 3$ and $M(3) \geq 12$. In 1954 N.F. Otrokov [262] proved that $M(n) \geq \frac{1}{2}(n^2 + 5n - 14)$ for $n \geq 6$ even, and $M(n) \geq \frac{1}{2}(n^2 + 5n - 26)$ for $n \geq 7$ odd. In 2015, for small degree systems, these values were improved by H. Liang & J. Torregrosa in [216] proving that $M(n) \geq n^2 + n - 2$ for $4 \leq n \leq 13$ perturbing holomorphic centers and using a new idea via parallelized computations. Taking into account the number of free parameters, $n^2 + 3n + 2$, in [121] it is conjectured that $M(n) = n^2 + 3n - 6$ for $n \geq 3$. This value is updated by one from the one previously conjectured in 2012 by J. Giné [119, 120]. The new conjecture removes only eight to the total number of parameters. Six corresponding to an affine change of variables that writes the linear part in its normal form, one corresponding to a rotation, and another to a rescaling. The previous conjecture took into account that the number of limit cycles in a center component does not change. But this is only generically. In fact, this is the key point of the proof of $M(3) \geq 12$, because the perturbed center has a 1-parameter rational first integral and the local cyclicity changes with this special parameter. Up to our knowledge, the best values for the local cyclicity for systems of degree four and five say that $M(4) \geq 21$ and $M(5) \geq 33$. These values, which are very close to the conjectured one, were provided in 2020, respectively, in [121] and [127].

- The previously commented local study of Otrokov provided also lower bounds for $\mathcal{H}(n)$ up to 1995. C. Christopher & N.G. Lloyd in [57] proved that $\mathcal{H}(n) \geq k\,n^2 \ln n$ for some constant k. In this result, the limit cycles surround many singularities. In 2003 J. Li improved this result in the survey [203], and proved that $\mathcal{H}(n) \geq \frac{1}{4}(n+1)^2\big(1.442695 \ln(n+1) - \frac{1}{6}\big) + n - \frac{2}{3}$.

In 2012 M. Han & J. Li gave a stronger result as follows.

Theorem 1.3 ([134]). *For any sufficiently small $\varepsilon > 0$ there exists a positive number n^*, depending on ε, such that*

$$\mathcal{H}(n) > \left(\frac{1}{2\ln 2} - \varepsilon \right)(n+2)^2 \ln(n+2), \quad for\ n > n^*.$$

Hence $\mathcal{H}(n)$ grows at least as rapidly as $\frac{1}{2\ln 2}(n+2)^2 \ln(n+2)$ for large n.

Moreover, they found out the lower bounds of $\mathcal{H}(n)$ for $n = 7, 8, \ldots, 30$, among them $\mathcal{H}(7) \geq 65$, $\mathcal{H}(8) \geq 67$, $\mathcal{H}(9) \geq 98$, $\mathcal{H}(10) \geq 100, \ldots$, $\mathcal{H}(30) \geq 1070$.

A lower bound with the same order has been obtained by M.J. Álvarez, B. Coll, P. De Maesschalck & R. Prohens in 2020 in [1] using slow-fast systems.

For small values of the degree n, in 2015, H. Liang & J. Torregrosa [216], obtained that $\mathcal{H}(6) \geq 40, \mathcal{H}(8) \geq 70$, and $\mathcal{H}(10) \geq 108$. In 2019, by using simultaneous degenerate Hopf bifurcations from 3 singularities, R. Prohens & J. Torregrosa [285] proved that $\mathcal{H}(4) \geq 28$, $\mathcal{H}(5) \geq 37$, $\mathcal{H}(6) \geq 53$, $\mathcal{H}(7) \geq 74$, $\mathcal{H}(8) \geq 96$, $\mathcal{H}(9) \geq 120$, $\mathcal{H}(10) \geq 142$, $\mathcal{H}(13) \geq 212$, $\mathcal{H}(17) \geq 384$, $\mathcal{H}(21) \geq 568$, $\mathcal{H}(31) \geq 1184$, $\mathcal{H}(35) \geq 1536$, $\mathcal{H}(39) \geq 1920$, and $\mathcal{H}(43) \geq 2272$.

1.1.5 Some results on Liénard equations

- For the generalized Liénard equation $\ddot{x} + f(x)\dot{x} + g(x) = 0$ (or equivalently, the planar systems $\dot{x} = y - F(x)$, $\dot{y} = -g(x)$, where $F(x) = \int_0^x f(s)\,ds$), Z. Zhang [371, 372] proved a theorem in 1958 that if the quotient $f(x)/g(x)$ is monotone, among some basic assumptions, then the limit cycle (if it exists) is unique. In particular, if $g(x) = x$, F is a cubic polynomial, and the system may have limit cycles, then $f(x)/x$ is monotone, so the corresponding Liénard equation has no more than one limit cycle, see the appendix of [184]. This theorem and different forms of its generalization were used widely, for example in [41, 342, 353, 366, 367]. Note that if a quadratic system is transformed to a Liénard equation, the functions F and g, in general, are no longer polynomials.

- Concerning the number of limit cycles for a polynomial Liénard equation (i.e. when F and g are polynomials), there is the so-called *Lins–De Melo–Pugh conjecture* in [218]: "if $g(x) = x$ and $\deg(F) = 2n + 1$ or $2n + 2$ $(n \geq 1)$, then the maximal number of limit cycles is n", see also Problem 7 of [311]. This conjecture was proved when $\deg(F) = 3$ in the same paper [218]. Note that this result can be proved in a very simple way by the theorem of [371] as mentioned above, see also the appendix of [184]. A counterexample to this conjecture was found in 2007 by F. Dumortier, D. Panazzolo & R. Roussarie [83] when the degree of F is seven. In 2011 P. De Maesschalck & F. Dumortier [70] gave counterexamples to this conjecture for $\deg(F) \geq 6$, see also [71]. All above counterexamples were constructed by using singular perturbations. In 2012 C. Li & J. Llibre [190] proved that the conjecture is true also for $\deg(F) = 4$. Recently, J. Llibre & X. Zhang have written a survey summarizing all these results, see [245]. The conjecture is still open only when the degree of the polynomial F is five.

- Concerning the number of small amplitude limit cycles in Liénard systems there is a series of works by N.G. Lloyd, C. Christopher, and S. Lynch, see for example [56, 57, 58, 247, 248], and by Y. Liu and J. Li, see [225, 226, 229]. Related to this topic, there are a lot of works by J. Llibre, A. Gasull and the research group in Barcelona, see for example [63, 98, 106, 107, 126, 240].

- In 2020 M.J. Álvarez, B. Coll, P. De Maesschalck & R. Prohens [1] prove, using slow-fast systems, that the best general lower bound in the Liénard class is of order $n \ln(n)/\ln(2)$. From it, they also provided the one for $\mathcal{H}(n)$ of order $n^2 \ln(n)/\ln(4)$ commented above.

For more details about limit cycles and Hilbert's 16th problem, we refer to the survey papers [24, 52, 67, 167, 203, 237, 302], and the books [3, 4, 53, 73, 130, 132, 169, 171, 214, 235, 251, 287, 293, 301, 354, 355, 373].

1.2   Weak Hilbert's 16th Problem

Now we turn to a weak version of the problem. Let $H = H(x, y)$ be a polynomial in x, y of degree $m \geq 2$ such that the level curves $\gamma_h \subset \{(x, y) : H(x, y) = h\}$ form a continuous family of ovals $\{\gamma_h\}$ for $h_1 < h < h_2$. Consider a polynomial 1-form $\omega = f(x, y)dy - g(x, y)dx$, where $\max(\deg(f), \deg(g)) = n \geq 2$. V.I. Arnold in [5, 6] proposed the following problem:

For fixed integers m and n find the maximum $Z(m, n)$ of the numbers of isolated zeros of the Abelian integrals

$$I(h) = \oint_{\gamma_h} \omega. \tag{1.2}$$

Recall that an *Abelian integral* is the integral of a rational 1-form along an algebraic oval. Note that in the above problem one must consider all possible H with all possible families of ovals $\{\gamma_h\}$, and arbitrary f and g. So it does not matter if we put $-$ or $+$ before g in ω. Remark also that the function $I(h)$ may be multivalued since it is possible that several ovals lie on the same level curve $H^{-1}(h)$.

At a first look, this problem has no relation with Hilbert's 16th problem at all. We will explain in Section 2.1 how these two problems are related to each other. Roughly speaking, the function $I(h)$, given by the Abelian integral (1.2), is the first approximation in ε of the *displacement function* of the Poincaré map on a segment transversal to γ_h (at least locally) for the system

$$X_{H,\varepsilon} : \quad \dot{x} = -\frac{\partial H(x, y)}{\partial y} + \varepsilon f(x, y), \quad \dot{y} = \frac{\partial H(x, y)}{\partial x} + \varepsilon g(x, y), \tag{1.3}$$

where H, f, and g are the same as above when defining the Abelian integral $I(h)$. Hence the number of isolated zeros of $I(h)$ (taking into account the multiplicities) gives an upper bound of the number of limit cycles of system (1.3) being ε small enough.

It is clear that if one takes $m = n + 1$, then system (1.3) is a special form of system (1.1), close to Hamiltonian for small ε. In this sense, the second problem usually is called the *weak (or tangential, infinitesimal)* Hilbert's 16th problem, and the number $\tilde{Z}(n) = Z(n+1, n)$ can be chosen as a lower bound of the Hilbert number $\mathcal{H}(n)$.

Independently, A.G. Khovanskii and A.N. Varchenko proved that for given m and n the number $Z(m,n)$ is uniformly bounded with respect to the choice of the polynomial H, the family of ovals $\{\gamma_h\}$, and the 1-form ω.

Theorem 1.4 ([174, 327]). $Z(m,n) < \infty$.

This result certainly is important. However, it is a purely existential statement, giving no information on the number $Z(m,n)$. To find an explicit expression for $Z(m,n)$ in general, even to find an explicit bound to it, is extremely hard. There are many works dealing with restricted versions of the problem (restriction on H or ω), some of them will be briefly introduced in these notes. It is natural to think about the possibility to find $\tilde{Z}(n) = Z(n+1,n)$ for lower n, and this was done by several authors over a period of about 10 years and only for $n = 2$. We will first introduce this result in the next subsection, then we will give more detailed information about a unified proof in Chapter 4.

For general n, G. Binyamini, D. Novikov & S. Yakovenko give a double exponential type upper bound for $\tilde{Z}(n)$ as follows.

Theorem 1.5 (Theorem 2 of [13]). $\tilde{Z}(n) < 2^{2^{\mathrm{Poly}(n)}}$, were the expression $\mathrm{Poly}(n) = O(n^p)$ stands for an explicit polynomially growing term with the exponent p not exceeding 61.

1.2.1 The study of $\tilde{Z}(2)$

We consider all cubic polynomials $H(x,y)$ with a continuous family of ovals $\{\gamma_h\}$ in $H^{-1}(h)$ for $h_c < h < h_s$, where h_c and h_s correspond to the critical values of the corresponding unperturbed quadratic Hamiltonian system X_H (i.e. (1.3) for $\varepsilon = 0$) at a center and a saddle loop respectively (the discussion below and Figure 1 show that this is the case for generic quadratic Hamiltonian systems). The family of ovals forms a *period annulus*. We first give a definition of the *cyclicity of the period annulus*.

Definition 1.6. For $0 < \sigma < (h_s - h_c)/2$, let N_σ be the maximum number of limit cycles of $X_{H,\varepsilon}$ which bifurcate from $\{\gamma_h : h \in [h_c + \sigma, h_s - \sigma]\}$ by quadratic perturbations. The *cyclicity of the period annulus* of X_H under quadratic perturbations is $\sup_\sigma N_\sigma$.

Note that by Theorem 1.4 the numbers N_σ are uniformly bounded in $\{\sigma\}$, hence the cyclicity is a finite number. For a precise meaning of limit cycles bifurcating from an oval γ_h of X_H, see Definition 2.3.

Recall that quadratic systems with at least one center are always integrable. They can be classified into the following five classes: Hamiltonian (Q_3^H), reversible (Q_3^R), generalized Lotka–Volterra (Q_3^{LV}), co-dimension 4 (Q_4), and the Hamiltonian triangle. This classification follows from [159] by using the terminology in [393], see also [300].

Definition 1.7 ([159])**.** A quadratic integrable system is said to be *generic* if it belongs to one of the first four classes and does not belong to other integrable ones. Otherwise, it is called *degenerate*.

It is shown by I.D. Iliev in [159] that if $X_H \in Q_3^H$ is generic, then the number $\tilde{Z}(2)$ gives the cyclicity of the period annulus of X_H. E. Horozov & I.D. Iliev prove in [154] that any cubic Hamiltonian, with at least one period annulus contained in its level curves, can be transformed into the following normal form,

$$H(x,y) = \frac{1}{2}(x^2 + y^2) - \frac{1}{3}x^3 + axy^2 + \frac{1}{3}by^3,$$

where a, b are parameters lying in the region

$$\bar{G} = \left\{ (a,b) : -\frac{1}{2} \le a \le 1,\ 0 \le b \le (1-a)\sqrt{1+2a} \right\}, \qquad (1.4)$$

and moreover, their respective vector fields X_H are generic if $(a,b) \in G = \bar{G} \setminus \partial\bar{G}$ and degenerate if $X_H \in \partial\bar{G}$.

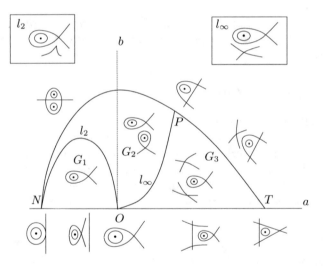

Figure 1: The phase portraits of X_H in Q_3^H.

Figure 1 shows all possible phase portraits of X_H for different ranges of a and b, where G is divided into three regions G_1, G_2, and G_3 by two curves l_2 and l_∞ (l_2 and l_∞ also belong to G). Along l_2 two singularities of X_H coincide, and when (a,b) tends to l_∞ a finite singularity of X_H coalesces with an infinite singularity. Hence, besides the two critical situations along l_2 and l_∞, X_H has one, two, or three saddle points if $(a,b) \in G_1, G_2$, or G_3, respectively. Moreover, X_H has two period annuli if $(a,b) \in G_2$ and on the arc PN over $\partial\bar{G}$, and one period annulus in the other cases.

Theorem 1.8. $\tilde{Z}(2) = 2$ *for generic cases.*

E. Horozov & I.D. Iliev [154] prove that the least upper bound of the number of zeros of related Abelian integrals is 2 for $(a,b) \in G_3$, later L. Gavrilov [113] obtains the same conclusion for $(a,b) \in G_1 \cup G_2$ (the method is also valid for $(a,b) \in G_3$). Since a basic assumption in [113] and [154] is that $H(x,y)$ has four distinct critical values (in the complex plane), the cases $(a,b) \in l_2 \cup l_\infty$ must be considered separately. Papers [258] and [375] independently give different proofs for $(a,b) \in l_\infty$, and [199] proves the same conclusion for the last case $(a,b) \in l_2$. A unified proof for all generic cases appears in [32].

Remark 1.9. If X_H is degenerate (i.e. $(a,b) \in \partial\bar{G}$), then it is not difficult to show that $I(h)$ has at most one zero. But this gives no information about the cyclicity of the period annulus, higher approximations must be considered. I.D. Iliev in [159] gives formulas (from the second- or third-order Melnikov functions, which will be discussed in Chapter 2) to determine the cyclicity for all degenerate cases. The cyclicity of the period annulus (or annuli) is 3 for the Hamiltonian triangle case ([156]), and is 2 for all other seven cases (see [54, 114, 157, 385, 389] and a unified proof in [188]).

Remark 1.10. X_H has two period annuli when $(a,b) \in G_2$. To prove that $\tilde{Z}(2)$ still is 2 in this case, implying that only $(1,1)$-distribution of limit cycles is possible if there are two nests of limit cycles after perturbation, [113] uses a result in [365] while [32] gives a direct proof. There is a similar study in [54] when the period annulus is bounded by the elliptic-segment loops.

Remark 1.11. The maximal number of zeros of the Abelian integral gives the maximal number of limit cycles of the perturbed system $X_{H,\varepsilon}$ bifurcating from the ovals in $\{\gamma_h : h \in (h_c, h_s)\}$ (see Theorem 2.4 (iv) on page 145 for a precise statement). If we would like to study the number of limit cycles that bifurcate from the whole annulus (or annuli), we also need to consider the ones bifurcating from the center, from the polycycle (homoclinic or heteroclinic loop), and from infinity, see Remark 2.6. This question for quadratic systems is still open, only some partial results appear in [115, 195].

1.2.2 Perturbations of elliptic and hyperelliptic Hamiltonians

Now we restrict the function H to the following form:

$$H(x,y) = \frac{y^2}{2} + P_m(x), \tag{1.5}$$

where P_m is a polynomial in x of degree m. The level curves of H are rational for $m = 1, 2$, elliptic for $m = 3, 4$, and hyperelliptic for $m \geq 5$. We assume $m \geq 2$ since the level curves have no ovals if $m = 1$.

We first give a general result about the main structure that have the Abelian integrals in this family of Hamiltonians.

Lemma 1.12. *Suppose that for the function H defined in (1.5) there is a family of ovals $\gamma_h \subset H^{-1}(h)$, and ω is an arbitrary polynomial 1-form of degree n, then*

$$\oint_{\gamma_h} \omega = \begin{cases} \oint_{\gamma_h} p_1(x)\, y\, dx, & n = 2, \\ \oint_{\gamma_h} p_k(x,h)\, y\, dx, & n \geq 3, \end{cases}$$

where p_1 is a linear function in x, and $p_k(x,h)$ is a polynomial in x and h of degree $k = m(n-1)/2$ if n is odd and $k = m(n-2)/2 + 1$ if n is even.

Proof. For every integers $i, j \geq 0$ it is easy to see that

$$\oint_{\gamma_h} x^i y^j\, dy = \begin{cases} 0, & i = 0, \\ -\frac{i}{j+1} \oint_{\gamma_h} x^{i-1} y^{j+1}\, dx, & i \geq 1. \end{cases}$$

Hence, without loss of generality we only consider $\omega = f(x,y)dx$, where f is a polynomial in x and y of degree n. On the other hand, we have

$$\oint_{\gamma_h} x^i y^j\, dx = \begin{cases} 0, & j = 2l, \\ \oint_{\gamma_h} x^i\, [2(h - P_m(x))]^l y\, dx, & j = 2l + 1. \end{cases}$$

The statements of the lemma immediately follow. $\square$

(A) The case $m = 2$.

In this case, we may choose $H = (x^2 + y^2)/2$ (to put the center of X_H at the origin) and the ovals are circles $\{x^2 + y^2 = h^2\}$. Suppose that the 1-form ω is of degree n, then by using the polar coordinates one finds that

$$\oint_{\gamma_{h^2}} \omega = h^2 Q_{n-1}(h),$$

where $Q_{n-1}(h)$ is a polynomial in h of degree $n - 1$, but depends only on h^2 by symmetry. Hence, $I(h)$ has at most $[(n-1)/2]$ zeros except the trivial zero at $h = 0$, which corresponds to the singularity at the origin.

(B) The elliptic Hamiltonian of degree 3.

In this case, if we suppose that the level curves of H contain a continuous family of ovals, then the two singularities of the corresponding vector field X_H must be a center and a saddle, which are chosen (without loss of generality) at $(-1, 0)$ and $(1, 0)$, respectively, and the elliptic Hamiltonian reads as

$$H(x,y) = \frac{y^2}{2} - \frac{x^3}{3} + x. \tag{1.6}$$

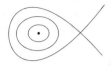

Figure 2: The family of ovals for case $m = 3$ in (1.5).

In this case the continuous family of ovals is given by

$$\{\gamma_h\} = \{(x, y) : H(x, y) = h, \ -2/3 \le h \le 2/3\}, \tag{1.7}$$

see Figure 2. By Lemma 1.12 the Abelian integral $I(h)$ can be expressed in the form $\oint_{\gamma_h} p_k(x, h) y \, dx$, where p_k is a polynomial in x and h. An important observation is that along γ_h we have

$$0 \equiv dH = H_x dx + H_y dy = (1 - x^2) dx + y \, dy,$$

which implies, also along γ_h, that $(1 - x^2) y \, dx + y^2 dy \equiv 0$, hence $I_2(h) \equiv I_0(h)$, where we define $I_j(h) = \oint_{\gamma_h} x^j y \, dx$. Similarly, we have

$$\oint_{\gamma_h} x^k (x^2 - 1) y \, dx = \oint_{\gamma_h} x^k y^2 dy = \oint_{\gamma_h} x^k (2h + 2x^3/3 - 2x) dy.$$

Using integration by parts on the right-hand side we find the following induction formula,

$$(2k + 9) I_{k+2}(h) - 3(2k + 3) I_k(h) + 6kh I_{k-1}(h) = 0,$$

where $k \ge 1$. Hence, it is not hard to prove the following result.

Lemma 1.13 ([274]). *Suppose that $I(h)$ is the Abelian integral of the polynomial 1-form ω of degree at most n over the ovals γ_h defined in (1.7), then*

$$I(h) = Q_0(h) I_0(h) + Q_1(h) I_1(h),$$

where Q_0 and Q_1 are polynomials with $\deg Q_0 \le [\frac{n-1}{2}]$ and $\deg Q_1 \le [\frac{n}{2}] - 1$. As usual $[\xi]$ means the integer part of ξ.

If we denote by $[(n-1)/2] = n_0$ and $[n/2] - 1 = n_1$, then $n_0 + n_1 = n - 2$, and any $I(h)$, defined in Lemma 1.13, can be expressed as a linear combination of the n independent functions

$$I_0(h), h I_0(h), h^2 I_0(h), \ldots, h^{n_0} I_0(h), I_1(h), h I_1(h), h^2 I_1(h), \ldots, h^{n_1} I_1(h).$$

Hence, it is possible to find a special $I(h)$, having $n - 1$ zeros for $h \in (-2/3, 2/3)$. On the other hand, one may expect that all such $I(h)$ have at most $n - 1$ zeros, counting their multiplicities. This result was proved by G.S. Petrov in [276]. Before stating his result we recall a definition.

Definition 1.14 ([256])**.** A $(k+1)$-tuple of smooth functions $(J_0, \ldots, J_k)$ defined on some open interval (h_0, h_1), is a *Chebyshev system*, if for any $\ell \leq k$, a nontrivial linear combination of the $\ell + 1$ functions $(J_0, \ldots, J_\ell)$ has at most ℓ zeros in the open interval (h_0, h_1), counting their multiplicities.

As we will see in Chapter 3, more concretely in Definition 3.8, we say that a set of $k + 1$ functions satisfying the above property, as the zeros have taken into account their multiplicities, is an *Extended Complete Chebyshev system* (ECT-system in short). An equivalent and simpler way to check the last statement in the above definition is the next result. See more details in [173].

Proposition 1.15. *A $(k+1)$-tuple of smooth functions $(J_0, \ldots, J_k)$ defined on some open interval (h_0, h_1), is an* Extended Complete Chebyshev system *if and only if, for $\ell = 0, \ldots, k$, the Wronskian determinants*

$$
W[J_0, \ldots, J_\ell] := \begin{vmatrix} J_0 & \cdots & J_\ell \\ J_0' & \cdots & J_\ell' \\ \vdots & \ddots & \vdots \\ J_0^{(\ell)} & \cdots & J_\ell^{(\ell)} \end{vmatrix}
$$

are non-vanishing in (h_0, h_1).

The simplest example is the set of monomials $(1, x, x^2, \ldots, x^k)$, which is a Chebyshev system on any interval. Another very interesting set is $(\ln x, 1, x \ln x, x, x^2 \ln x, x^2, \ldots, x^k \ln x, x^k)$, which is a Chebyshev system on any positive interval. In fact, the above property is very useful to study versal unfoldings, because it provides the same bifurcation diagram for the zeros of any linear combination of $(J_0, \ldots, J_k)$ as the one obtained for the set of monomials $(1, x, x^2, \ldots, x^k)$.

Theorem 1.16 ([275, 276])**.** *The space of functions $\{I(h)\}$, defined in Lemma 1.13, has the Chebyshev property on $h \in (-2/3, 2/3)$. This means that any nontrivial $I(h)$ has at most $n - 1$ zeros, and there exists a 1-form ω, such that $I(h)$ has exactly $n - 1$ zeros.*

In fact, Petrov made an analytic extension of $I(h)$ from $(-2/3, 2/3)$ to a domain D in the complex plane, and proved by using the Argument Principle that the space of extended functions has also the Chebyshev property in D. We will detail this proof also in Chapter 3.

One motivation for studying the perturbations of the elliptic Hamiltonians comes from the so-called Bogdanov–Takens bifurcation, see [15, 317, 318]. If a $\mathcal{C}^\infty$ planar system has a nilpotent linear part, a truncated normal form up to degree 2 looks like

$$
\dot{x} = y, \quad \dot{y} = ax^2 + bxy. \tag{1.8}
$$

If $ab \neq 0$, then the problem has codimension 2 and, by a scaling if necessary, the parameters (a, b) can be assumed to be $(1, \pm 1)$. A universal unfolding (in $\mathcal{C}^\infty$ class) could be (see, for example, [53])

$$\dot{x} = y,$$
$$\dot{y} = \mu_1 + \mu_2 y + x^2 + xyF(x,\mu) + y^2 G(x,y,\mu), \tag{1.9}$$

where $\mu = (\mu_1, \mu_2)$ are small parameters, $F, G \in C^\infty$, and $F(0,0) = \pm 1 = \mathrm{sgn}(ab)$. There is no bifurcation for $\mu_1 > 0$ and the saddle-node bifurcation happens for $\mu_1 = 0$. The most interesting phenomenon appears for $\mu_1 < 0$, in this case by a change of coordinates and parameters

$$\mu_1 = -\varepsilon^4, \ \mu_2 = \alpha\varepsilon^2, \ x = \varepsilon^2 \bar{x}, \ y = \varepsilon^3 \bar{y}, \ t = \bar{t}/\varepsilon,$$

where $\varepsilon > 0$ small, system (1.9) (changing $(\bar{x}, \bar{y}, \bar{t})$ back to (x, y, t)) becomes

$$\dot{x} = y,$$
$$\dot{y} = -1 + x^2 + \varepsilon(\alpha \pm x)y + O(\varepsilon^2), \tag{1.10}$$

which is exactly the perturbation of X_H with H in the form (1.6), and the corresponding Abelian integral is

$$I(h) = \alpha I_0(h) \pm I_1(h). \tag{1.11}$$

As a typical example, we will introduce several methods to study the number of zeros of this Abelian integral in Chapter 3.

The quadratic normal form (1.10) cut up to first-order in ε was studied globally by L.M. Perko in [273] that, using an adequate scaling, it writes as

$$\dot{x} = y,$$
$$\dot{y} = -n + by + x^2 + xy. \tag{1.12}$$

This system has exactly one hyperbolic limit cycle if and only if $n > 0$ and $b^*(n) < b < \sqrt{n}$. The function $b^* : \mathbb{R}^+ \to \mathbb{R}$ defines the curve $b = b^*(n)$, for $n \geq 0$, where the phase portrait (1.12) has a hyperbolic and unstable homoclinic loop. Moreover, a Hopf bifurcation occurs on the curve $b = \sqrt{n}$. More details on the bifurcation diagram can be seen in [53, 197]. The goal of Perko's work in 1992 is the proof of that the function b^* is analytic. He also identify the first term of its Taylor series expansion near the origin. In this paper he conjectured that this curve goes to infinity as $b^*(n) \approx \sqrt{n} - 1$. This conjecture was proved using the Bendixson-Dulac criterion in 2010 by H. Giacomini, A. Gasull & J. Torregrosa in [97] showing that the function b^* satisfies

$$\max\left(\frac{5}{7}\sqrt{n}, \sqrt{n} - 1\right) \leq b^*(n) < \min\left(\frac{5\sqrt{n} + \frac{37}{12}n}{7 + \frac{37}{12}\sqrt{n}}, \sqrt{n} - 1 + \frac{25}{7\sqrt{n}}\right).$$

An improvement of these lower and upper bounds can be found in [96]. The study near infinity for (1.12) is developed in [18].

If $b = 0$ in (1.8), then the problem has higher codimension; the study of the bifurcations for codimensions 3 and 4 (not only the study of zeros of the corresponding Abelian integrals, but also the number of limit cycles and the bifurcation diagrams) was given in [87] and [196], respectively.

(C) The elliptic Hamiltonian of degree 4.

In this case, we may take the function H in the form

$$H(x,y) = \frac{y^2}{2} + a\frac{x^4}{4} + b\frac{x^3}{3} + c\frac{x^2}{2}, \tag{1.13}$$

where $a \neq 0$. There are five types of continuous families of ovals on the level curves of H, shown in Figure 3 depending on the values of the parameters (a, b, c), called the truncated pendulum case, the saddle loop case, the global center case, the cuspidal loop case, and the figure-eight loop (Duffing oscillator) case, respectively.

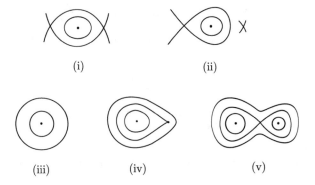

Figure 3: The families of ovals for the case $m = 4$ in (1.5).

The first two cases correspond to $a < 0$ while the last three cases correspond to $a > 0$. Note that in the figure-eight loop case the corresponding Abelian integral is a multi-valued function, since an oval in the left annulus and an oval in the right annulus (surrounded by the figure-eight loop) may correspond to a same value of h.

In [277] and [278] G.S. Petrov considered the figure-eight loop case ($a > 0$, $b < 0$, and $H(x,0)$ having only three real different critical values), and the case $(a, b, c) = (1, 1, 1)$, a special case of the global center, respectively. In the first case he obtained the result concerning the two annuli surrounded by the figure-eight loop. Recently C. Liu in [219] studied the region outside the figure-eight loop, and considered the total number of zeros for the ovals in the two annuli surrounded by the figure-eight loop. We state their results in the following theorems.

Theorem 1.17 ([277]). *Let H be as in (1.13) with the figure-eight loop. Then the space of the elliptic integrals $I(h)$ of a 1-form of degree n over cycles vanishing at one of the two singularities of X_H surrounded by the figure-eight loop has* the Chebyshev property on the corresponding intervals of h. This means that the number of zeros of nontrivial $I(h)$ is less than the dimension of the space. This dimension is $n + [(n-1)/2]$.

The conclusion of [278] for $(a, b, c) = (1, 1, 1)$ (a global center case) is similar, the dimension of the space of Abelian integrals in this case is $2[(n-1)/2] + 1$. We notice that in these results, $[\cdot]$ denotes the integer part function.

Theorem 1.18 ([219]). *Let H be as in (1.13) with the figure-eight loop and ω be a polynomial 1-form of degree n. Then the following statements hold.*

(i) *The total number of zeros of $I(h)$ (taking into account their multiplicities) for the ovals in the two annuli surrounded by the figure-eight loop does not exceed $2n - 1$ for n even, or $2n + 1$ for n odd.*

(ii) *The number of zeros (taking into account their multiplicities) of $I(h)$ for the ovals outside the figure-eight loop does not exceed $2n + 1$ for n even, or $2n + 3$ for n odd.*

There is a series of papers dealing with the exact number of zeros of the Abelian integrals over all types of ovals in Figure 3, but which only consider 1-forms of degree 3 as follows:

$$\omega = (\alpha + \beta x + \gamma x^2) \, y \, dx. \tag{1.14}$$

Comparing with the general 1-form of degree 3, one term, $y^3 dx$, is omitted. Let us first explain why the 1-form (1.14) is interesting. Consider a cubic Liénard equation with small quadratic damping:

$$\ddot{x} + \varepsilon \, p_2(x) \, \dot{x} + p_3(x) = 0,$$

where ε is a small parameter and p_k are polynomials in x of degree k. This equation is equivalent to the planar system

$$\dot{x} = y, \quad \dot{y} = -p_3(x) - \varepsilon p_2(x) \, y. \tag{1.15}$$

The study of the number of limit cycles of system (1.15) naturally leads to the study of the Abelian integral of the 1-form (1.14) over the ovals of (1.13).

Theorem 1.19. *Let $I(h)$ be an Abelian integral of the polynomial 1-form (1.14) over the ovals contained in the level curves of the elliptic Hamiltonian of degree 4 (1.13). Then the maximal number of zeros of $I(h)$ (taking into account their multiplicities) is*

(i) *1 in the truncated pendulum case ([151]);*

(ii) *2 in the saddle loop case ([78]);*

(iii) *4 in the global center case, and the four zeros of the elliptic integral can be simple or multiple exhibiting a complete unfolding of a zero of multiplicity 4 ([79]);*

(iv) 4 *in the cuspidal loop case, moreover, if restricting to the level curves "inside" and "outside" the cuspidal loop, we found the sharp upper bound to be, respectively, 2 and 3 ([80]);*

(v) 5 *in the figure-eight loop case, and there are three kinds of zeros for the elliptic integrals, depending on the integral over compact level curves inside the left loop, inside the right loop, or outside the figure-eight loop. We denote their respective numbers by* n_1, n_2, n_3, *then* $n_1+n_2 \leq 2$, $n_3 \leq 4$, *and* $n_1+n_2+n_3 \leq 5$ *(see [81] for a precise description).*

Remark 1.20. The results in Theorem 1.19 are valid for $b \neq 0$. If $b = 0$, then the Hamiltonians (1.13) are symmetric (called also reversible); this may happen for the cases (i), (iii), and (v) of Figure 3. If $0 < b \ll 1$, the parameter b breaks this symmetry in a generic way, one has to add it into the parameter space of perturbations, and for a description of the bifurcation diagram of the unfolding in a full neighborhood of the origin in the parameter space, see [163, 194].

Remark 1.21. The results of Theorem 1.19 give an estimate of the number of limit cycles bifurcating from the annulus (annuli) up to first-order of approximation in ε, and for special perturbations up to degree 3 in (x, y). It is naturally to consider the same problem for any order of approximation in ε, and for general perturbations up to degree 3 in (x, y). For this purpose [118] studied the non-symmetric cases, and [162] studied one of symmetric cases, see Examples 2.18 and 2.19 in Section 2.2 for more details.

(D) The hyperelliptic case.

In this case the polynomial $P_m(x)$ in (1.5) has degree at least 5. To find the exact number of zeros of the Abelian integrals for small n or to give an explicit upper bound for the number of zeros in general, such as we have introduced above for $m = 3, 4$, is very hard. One upper bound can be found in Theorem 1.5 for general H, with a tower function (iterated exponent). We believe that this bound is much bigger than the sharp one.

There are some other works ([170, 260], for example) which give an explicit upper bound of the number $Z(m, n)$ by certain tower functions, not restricted to the hyperelliptic case, but with other restrictions on the Hamiltonians.

The following result is about a lower bound for $\tilde{Z}(n) = Z(n + 1, n)$.

Theorem 1.22 ([164]). *If* $H \in \mathbb{R}[x, y]$ *is a Morse polynomial of degree* $n+1$ *transversal to infinity, then for any* $N = \frac{1}{2}(n + 1)(n - 2)$ *real ovals* $\{\gamma_h \subset H^{-1}(h)\}$ *on* $\mathbb{R}^2$ *one can construct a 1-form* $\omega = P(x, y)dx + Q(x, y)dy$, *being* P, Q *real polynomials in* (x, y) *with* $\deg P, \deg Q \leq n$, *such that the perturbation* $\{dH + \varepsilon\omega = 0\}$ *produces at least* N *limit cycles which converge to the specified ovals as* $\varepsilon \to 0$.

A *Morse function* means that all its critical points are non-degenerate, and all critical values are different (see [7], for example); a polynomial $f \in \mathbb{C}[x, y]$ of degree $n + 1 \geq 2$ is called *transversal to infinity*, if one of the two equivalent conditions holds:

(i) Its principal homogeneous part factors out as the product of $n + 1$ pairwise different linear forms.

(ii) Its principal homogeneous part has an isolated critical point of multiplicity n^2 at the origin.

Chapter 2

Abelian Integrals and Limit Cycles

In this chapter we will explain the relation between the number of zeros of the Abelian integrals and the number of limit cycles of the corresponding planar polynomial differential systems.

2.1 Poincaré–Pontryagin Theorem

Now we consider a polynomial $H(x,y)$ of degree m as in the previous chapter, the corresponding Hamiltonian vector field X_H,

$$\frac{dx}{dt} = -\frac{\partial H(x,y)}{\partial y}, \quad \frac{dy}{dt} = \frac{\partial H(x,y)}{\partial x}, \tag{2.1}$$

and a perturbed system

$$\frac{dx}{dt} = -\frac{\partial H(x,y)}{\partial y} + \varepsilon f(x,y), \quad \frac{dy}{dt} = \frac{\partial H(x,y)}{\partial x} + \varepsilon g(x,y), \tag{2.2}$$

where f and g are polynomials in x,y of degrees at most n, and ε is a small parameter.

Suppose that there is a family of ovals, $\gamma_h \subset H^{-1}(h)$, continuously and monotonically depending on a level $h \in (a,b)$. Then we may define the Abelian integral, as in the previous chapter, as

$$I(h) = \oint_{\gamma_h} f(x,y)dy - g(x,y)dx. \tag{2.3}$$

It is clear that all γ_h, filling up an annulus for $h \in (a,b)$, are periodic orbits of the unperturbed Hamiltonian system (2.1).

Consider the question: *How many orbits keep being unbroken and become the periodic orbits of the perturbed system (2.2) for small ε?* Note that if the number of such orbits is finite, then they are limit cycles of (2.2).

This question can be proposed in the converse way: Is it possible to find a value $h \in (a, b)$, and some periodic orbits Γ_ε of the perturbed system (2.2), such that Γ_ε tends to γ_h (in the sense of Hausdorff distance) as ε goes to 0? And, how many of such Γ_ε for the same h?

To answer this question, we take a segment σ, transversal to each oval γ_h. We choose the values of the function H itself to parameterize σ, and denote by $\gamma_h(\varepsilon)$ a piece of the orbit of the perturbed system (2.2) between the starting point h on σ and the next intersection point $\Pi(h, \varepsilon)$ with σ, see Figure 4. The "next

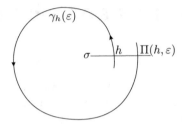

Figure 4: Construction of the displacement function.

intersection" is possible for sufficiently small ε, since $\gamma_h(\varepsilon)$ is close to γ_h. As usual, the difference $d(h, \varepsilon) = \Pi(h, \varepsilon) - h$ is called the *displacement function* and $\Pi(h, \varepsilon)$ the Poincaré return map.

Theorem 2.1 (Poincaré–Pontryagin [280, 281]). *We have that*

$$d(h, \varepsilon) = \varepsilon \left(I(h) + \varepsilon\phi(h, \varepsilon) \right), \ as \ \varepsilon \to 0, \tag{2.4}$$

where $\phi(h, \varepsilon)$ is analytic and uniformly bounded for (h, ε) in a compact region near $(h, 0)$ for all $h \in (a, b)$.

Proof. By the construction above, the displacement function is given by the difference of the function H between the two endpoints of $\gamma_h(\varepsilon)$, that is

$$d(h, \varepsilon) = \int_{\gamma_h(\varepsilon)} dH = \int_{\gamma_h(\varepsilon)} \left(\frac{\partial H}{\partial x} \frac{dx}{dt} + \frac{\partial H}{\partial y} \frac{dy}{dt} \right)\bigg|_{(2.2)} dt.$$

Substituting (2.2) into the right-hand side, we find

$$d(h, \varepsilon) = \varepsilon \int_{\gamma_h(\varepsilon)} \left(\frac{\partial H}{\partial x} f + \frac{\partial H}{\partial y} g \right)\bigg|_{(2.2)} dt.$$

Note that $\gamma_h(\varepsilon)$ converges to γ_h uniformly as ε goes to 0 since γ_h is compact, and $H_x dt = dy$, $H_y dt = -dx$ along γ_h by (2.1), we immediately obtain (2.4), where $I(h)$ is given by (2.3). ☐

Remark 2.2. Note that the number of zeros of the displacement function is independent of the choice of the transversal segment σ.

From Theorem 2.1 we obtain the following result giving an answer to the above question. We use X_H and $X_{H,\varepsilon}$ to denote the Hamiltonian system (2.1) and its perturbation (2.2) respectively, and first give a definition for convenience.

Definition 2.3. If there exist an $h^* \in (a,b)$ and an $\varepsilon^* > 0$ such that $X_{H,\varepsilon}$ has a limit cycle Γ_ε for $0 < |\varepsilon| < \varepsilon^*$, and Γ_ε tends to γ_{h^*} as $\varepsilon \to 0$, then we will say that Γ_ε *bifurcates from* γ_{h^*}. We say that a limit cycle Γ of $X_{H,\varepsilon}$ *bifurcates from the annulus* $\cup_{h \in (a,b)}\gamma_h$ of X_H, if there is an $h \in (a,b)$ such that Γ bifurcates from γ_h.

Theorem 2.4. *We suppose that the Abelian integral $I(h)$ is not identically zero for $h \in (a,b)$, then the following statements hold.*

(i) *If $X_{H,\varepsilon}$ has a limit cycle bifurcating from γ_{h^*}, then $I(h^*) = 0$.*

(ii) *If there exists an $h^* \in (a,b)$ such that $I(h^*) = 0$ and $I'(h^*) \neq 0$, then $X_{H,\varepsilon}$ has a unique limit cycle bifurcating from γ_{h^*}. Moreover, this limit cycle is hyperbolic.*

(iii) *If there exists an $h^* \in (a,b)$ such that $I(h^*) = I'(h^*) = \cdots = I^{(k-1)}(h^*) = 0$, and $I^{(k)}(h^*) \neq 0$, then $X_{H,\varepsilon}$ has at most k limit cycles bifurcating from the same γ_{h^*}, taking into account the multiplicities of the limit cycles.*

(iv) *Up to the first-order approximation in ε, the total number of limit cycles (counting their multiplicities) of $X_{H,\varepsilon}$, bifurcating from the annulus $\cup_{h \in (a,b)}\gamma_h$ of X_H, is bounded by the maximum number of isolated zeros (counting their multiplicities) of the Abelian integral $I(h)$ for $h \in (a,b)$.*

Proof. (i) Suppose that a limit cycle Γ_ε of $X_{H,\varepsilon}$ bifurcates from γ_{h^*}. By Theorem 2.1, there exist an $\varepsilon^* > 0$ and $h_\varepsilon \to h^*$ as $\varepsilon \to 0$, such that

$$d(h_\varepsilon, \varepsilon) = \varepsilon(I(h_\varepsilon) + \varepsilon\phi(h_\varepsilon, \varepsilon)) \equiv 0, \quad 0 < |\varepsilon| < \varepsilon^*.$$

Dividing by ε on both sides, and taking the limit as $\varepsilon \to 0$, we obtain $I(h^*) = 0$.

(ii) Suppose that there exists an $h^* \in (a,b)$ such that $I(h^*) = 0$ and $I'(h^*) \neq 0$. Since we consider limit cycles for small ε and $\varepsilon \neq 0$, instead of the displacement function $d(h,\varepsilon)$ we may study the zeros of $\tilde{d}(h,\varepsilon) = d(h,\varepsilon)/\varepsilon$. By Theorem 2.1 we have

$$\tilde{d}(h,\varepsilon) = I(h) + \varepsilon\phi(h,\varepsilon),$$

where ϕ is analytic and uniformly bounded in a compact region near $(h^*,0)$. Since $\tilde{d}(h^*,0) = I(h^*) = 0$ and $\tilde{d}'_h(h^*,0) = I'(h^*) \neq 0$, by the Implicit Function Theorem, we find an $\varepsilon^* > 0$, an $\eta^* > 0$, and a unique function $h = h(\varepsilon)$ defined in $U^* = \{(h,\varepsilon) : |h - h^*| \leq \eta^*, |\varepsilon| \leq \varepsilon^*\}$, such that $h(0) = h^*$ and $\tilde{d}(h(\varepsilon), \varepsilon) \equiv 0$, for $(h,\varepsilon) \in U^*$. Hence, the unique $h(\varepsilon)$ gives a unique limit cycle Γ_ε of system (2.2) for each small ε. Only remains to prove that the bifurcated limit cycle Γ_ε is hyperbolic (for small ε). This fact is easy to check because it comes from a simple zero of $I(h)$ at h^*. We give a precise proof below. We write $I(h)$ in (2.3) as $I_1(h) + I_2(h)$, where

$$I_1(h) = \oint_{\gamma_h} f(x, y)dy, \quad I_2(h) = -\oint_{\gamma_h} g(x, y)dx.$$

In the first integral we treat x as a function of y and h, and along γ_h we have $H_x x_h = 1$ and $dy = H_x dt$. This gives

$$I_1'(h) = \oint_{\gamma_h} f_x x_h dy = \oint_{\gamma_h} f_x dt.$$

Similarly we obtain

$$I_2'(h) = -\oint_{\gamma_h} g_y y_h dx = \oint_{\gamma_h} g_y dt.$$

Hence $\oint_{\gamma_{h^*}} (f_x + g_y)dt = I'(h^*) \neq 0$, which implies

$$\oint_{\Gamma_\varepsilon} \text{trace}\,(2.2)dt = \varepsilon \oint_{\Gamma_\varepsilon} (f_x + g_y)dt \neq 0, \quad 0 < |\varepsilon| \ll 1,$$

since $\Gamma_\varepsilon \to \gamma_{h^*}$ as $\varepsilon \to 0$. The hyperbolicity of Γ_ε follows.

(iii) Assume that there exists an $h^* \in (a, b)$ such that $I(h^*) = I'(h^*) = \cdots = I^{(k-1)}(h^*) = 0$, and $I^{(k)}(h^*) \neq 0$. We need to show that there exist a $\delta > 0$ and an $\eta > 0$, such that for any $(h, \varepsilon) \in U = \{|h - h^*| < \eta, |\varepsilon| < \delta\}$, the displacement function $d(h, \varepsilon)$ has at most k zeros in h, taking into account their multiplicities. Suppose the contrary, then for any integer j there exist $\varepsilon_j > 0$ and $\eta_j > 0$, $\varepsilon_j \to 0$ and $\eta_j \to 0$ as $j \to \infty$, such that for any ε_j the function $d(h, \varepsilon_j)/\varepsilon_j$ has at least $k + 1$ zeros for $|h - h^*| < \eta_j$. By using the Rolle's Theorem we find an h_j such that $|h_j - h^*| < \eta_j$ and

$$I^{(k)}(h_j) + \varepsilon_j \frac{\partial^k}{\partial^k h} \phi(h_j, \varepsilon_j) = 0,$$

which implies $I^{(k)}(h^*) = 0$ by taking the limit as $j \to \infty$, leading to a contradiction.

(iv) This statement is a consequence of the first three. In fact, for any small $\sigma > 0$ we may consider the number of limit cycles, up to first-order approximation in ε and bifurcating from $\{\gamma_h, h \in [a+\sigma, b-\sigma]\}$ for small ε. By Theorem 1.4 these numbers are uniformly bounded with respect to σ, and the maximum of these numbers for such $\{\sigma > 0\}$ is bounded by the maximum number of isolated zeros (counting their multiplicities) of the Abelian integral $I(h)$ for $h \in (a, b)$. $\square$

Example 2.5. Consider the van der Pol equation

$$\ddot{x} + \varepsilon(x^2 - 1)\dot{x} + x = 0,$$

which is equivalent to the system

$$\dot{x} = y, \quad \dot{y} = -x + \varepsilon(1 - x^2)y. \tag{2.5}$$

When $\varepsilon = 0$, equation (2.5) is a Hamiltonian system, $H(x,y) = x^2 + y^2$, with a family of ovals

$$\gamma_h = \{(x,y) : x^2 + y^2 = h^2, \ h > 0\}.$$

By using the polar coordinates $x = h\cos\theta, y = h\sin\theta$, and noting that the orientation of γ_h is clockwise, from formula (2.3) we have

$$I(h) = -\oint_{\gamma_h} (1 - x^2)y\,dx = \int_0^{2\pi} (1 - h^2\cos^2\theta)h^2(-\sin^2\theta)\,d\theta = \pi h^2\left(\frac{h^2}{4} - 1\right).$$

The zero $h = 0$ corresponds to the singularity of the system and $h = 2$ is the only positive zero of $I(h)$, which is simple because $I'(2) = 4\pi$. Using Theorem 2.4 we conclude that, for small ε, system (2.5) has a unique limit cycle which is hyperbolic and tends to the circle of radius 2 as ε goes to 0.

In fact, the limit cycle, for $\varepsilon \neq 0$, is always hyperbolic and unique. The simpler proof that (2.5) has at most one limit cycle uses the existence of a Dulac function $B(x,y) = (x^2 + y^2 - 1)^{-1/2}$. This nice function B was discovered by L.A. Cherkas. See more details, for example, in the textbook of C. Chicone [47]. □

Remark 2.6. Let us give a precise meaning of *cyclicity* of an annulus. In Theorem 2.4 we consider $I(h)$ for $h \in (a,b)$, the end points may correspond to critical values of H. For example, γ_h shrinks to a center of X_H if $h \to a$, and γ_h expands to a polycycle (homoclinic or heteroclinic orbit) or expands to infinity if $h \to b$. We suppose that the cyclicity of the annulus is M, then for any $\sigma > 0$, by the proof of (iii) above and the compactness of $[a + \sigma, b - \sigma]$, there is $\varepsilon_\sigma > 0$ such that for any $|\varepsilon| < \varepsilon_\sigma$, system $X_{H,\varepsilon}$ has at most M limit cycles (counting their multiplicities) bifurcating from $\{\gamma_h\}$ for $h \in [a + \sigma, b - \sigma]$. But if $\sigma \to 0$, it is possible that also $\varepsilon_\sigma \to 0$. Hence, one does not give a uniform estimate of the number of limit cycles bifurcating from $\{\gamma_h\}$ for $h \in [a,b]$, since at the values $h = a$ and $h = b$ the Implicit Function Theorem (or the Rolle's Theorem) can not be applied to the displacement function. It is well known that if the center of X_H (corresponding to $h = a$) is non-degenerate and $X_{H,\varepsilon}$ is analytic, then $I(h)$ can be extended at the value a analytically (see Theorem 3.9 of [214] or Lemma 20 of [26]), and the above conclusion can be extended to $[a,b)$. In general, it is hard to extend to $[a,b]$. It has been shown by an example with two-saddle loop in the paper [27] (also see [84, 135, 146, 250]) that some *Alien limit cycle(s)* can appear, that is the total number of limit cycles bifurcating from $\{\gamma_h\}$ for $h \in [a,b]$ is more than the maximal number of zeros of $I(h)$ for $h \in (a,b)$. On limit cycle bifurcation from a polycycle, see [53, 132] and the references therein.

Remark 2.7. If $H = H_\nu$ depends on some parameter ν, then for some special values, say ν^*, H_{ν^*} may be *degenerate*, for example has some symmetries, see the cases $X_H \in Q_3^H \cap \{Q_3^R \cup Q_3^{LV}\}$ (i.e. $(a,b) \in \partial G$ in Figure 2), or the cases (i), (iii), and (v) in Figure 3 for some special values of the parameters of H. As I.D. Iliev explained in [159] the degeneracy causes a lower bound for the number of zeros of $I(h)$ than the expected one, and the function $I(h)$ (even in the case that $I(h)$ is not identically zero) can never yield the maximum number of zeros of the displacement function $d(h, \varepsilon)$ for the whole class of perturbations. In this case a higher-order (in ε) approximation of $d(h, \varepsilon)$ is needed, as in the study of the cyclicity of the period annulus for $X_H \in Q_3^H \cap \{Q_3^R \cup Q_3^{LV}\}$ (see Remark 1.9); or the parameters which break down the symmetry of H and the perturbation parameters should be considered together in higher dimensional space to give a "principal part" of the displacement function, as in the study of perturbations of symmetric elliptic Hamiltonians of degree 4 in [194].

Remark 2.8. Theorems 2.1 and 2.4 were proved for the polynomial Hamiltonian systems (2.1) and their polynomial perturbation (2.2), but the proofs for analytic vector fields are essentially the same.

Remark 2.9. Theorem 2.4 was generalized to three dimensions under certain conditions in [193], and used to study the number of periodic orbits of a perturbed three-dimensional Volterra system.

2.2   Higher Order Approximations

It is shown in the last section that the Abelian integral $I(h)$, related to X_H, gives the first-order approximation of the displacement function of the perturbed system $X_{H,\varepsilon}$. Hence the number of isolated zeros of $I(h)$ gives an upper bound of the number of limit cycles of $X_{H,\varepsilon}$, if $I(h)$ is not identically zero. In this section, we continue the discussion of the problem if $I(h) \equiv 0$ for $h \in (a, b)$.

It is very natural to express the displacement function in the form

$$d(h, \varepsilon) = \varepsilon I_1(h) + \varepsilon^2 I_2(h) + \cdots + \varepsilon^j I_j(h) + O(\varepsilon^{j+1}), \tag{2.6}$$

where $I_1(h) \equiv I(h)$ for ε small. The question is that if $I_1(h) \equiv 0$, then how to compute the second-order approximation $I_2(h)$ and so on?

The following algorithm to compute $I_{k+1}(h)$, if $I_j(h) \equiv 0$ for $j = 1, 2, \ldots, k$, was given by J.-P. Françoise in [90], see also [344].

Denote $dH = H_x dx + H_y dy$ and $\omega = f dy - g dx$, where H, f, and g are polynomials in x and y, $\deg(H) = n + 1$, and $\max(\deg(f), \deg(g)) = n$. Then equations (2.1) and (2.2) can be written in Pfaffian forms $dH = 0$ and $dH - \varepsilon\omega = 0$, respectively. As before, we use γ_h to denote the family of ovals contained in the level curves $H^{-1}(h)$, σ a segment transversal to γ_h and parameterized by H, and $\gamma_h(\varepsilon)$ a piece of the orbit of $dH - \varepsilon\omega = 0$ between the starting point h on σ and the next intersection point $\Pi(h, \varepsilon)$ with σ. By using these notations the Theorem of Poincaré–Pontryagin can be shown in brief as follows.

The integration of $dH - \varepsilon\omega = 0$ over $\gamma_h(\varepsilon)$ gives

$$d(h, \varepsilon) = \int_{\gamma_h(\varepsilon)} dH = \varepsilon \int_{\gamma_h(\varepsilon)} \omega = \varepsilon \int_{\gamma_h} \omega + O(\varepsilon^2).$$

Following [90], we say that the polynomial H satisfies the condition $(*)$ if and only if for all polynomial 1-forms ω we have that:

$$\int_{\gamma_h} \omega = 0 \Leftrightarrow \text{there are polynomials } q \text{ and } R \text{ such that } \omega = q \, dH + dR. \qquad (*)$$

Theorem 2.10 ([90]). *Assume that H satisfies the condition $(*)$ and $I_j(h) \equiv 0$ (in the formula (2.6)) for $j = 1, 2, \ldots, k$. Then there are $q_1, \ldots, q_k$ and $R_1, \ldots, R_k$ such that $\omega = q_1 \, dH + dR_1, q_1\omega = q_2 \, dH + dR_2, \ldots, q_{k-1}\omega = q_k \, dH + dR_k$, and*

$$I_{k+1}(h) = \int_{\gamma_h} q_k \, \omega.$$

Proof. The proof can be done by induction.

(1) Assume that $I_1(h) \equiv 0$. By the condition $(*)$ we find two polynomials q_1 and R_1 such that $\omega = q_1 \, dH + dR_1$, which implies the next equality

$$(1 + \varepsilon q_1)(dH - \varepsilon \, \omega) = d\,(H - \varepsilon R_1) - \varepsilon^2 q_1 \, \omega.$$

After integrating the above equality over $\gamma_h(\varepsilon)$, along which $dH - \varepsilon\omega = 0$, we obtain that the displacement function writes as

$$d(h, \varepsilon) = \int_{\gamma_h(\varepsilon)} dH = \varepsilon^2 \int_{\gamma_h(\varepsilon)} q_1\omega + \varepsilon \int_{\gamma_h(\varepsilon)} dR_1.$$

Treating $dR_1 = \frac{\partial R_1}{\partial x} dx + \frac{\partial R_1}{\partial y} dy$ as a polynomial 1-form and using $I_1(h) \equiv 0$, we get $\int_{\gamma_h(\varepsilon)} dR_1 = O(\varepsilon^2)$. Thus we obtain

$$d(h, \varepsilon) = \varepsilon^2 \int_{\gamma_h(\varepsilon)} q_1\omega + O(\varepsilon^3) = \varepsilon^2 \int_{\gamma_h} q_1\omega + O(\varepsilon^3),$$

i.e. $I_2(h) = \int_{\gamma_h} q_1\omega$.

(2) Suppose that $q_{j-1}\omega = q_j \, dH + dR_j$ (denote $g_0 = 1$) and $I_j(h) \equiv 0$ for $j = 1, \ldots, k$. Then the former gives the equality

$$(1 + \varepsilon q_1 + \cdots + \varepsilon^k q_k)(dH - \varepsilon\,\omega) = d\,(H - \varepsilon R_1 - \cdots - \varepsilon^k R_k) - \varepsilon^{k+1} q_k \, \omega,$$

and the latter gives $\int_{\gamma_h(\varepsilon)} d\,(\varepsilon R_1 + \cdots + \varepsilon^k R_k) = O(\varepsilon^{k+2})$. Hence

$$d(h, \varepsilon) = \varepsilon^{k+1} \int_{\gamma_h(\varepsilon)} q_k\omega + O(\varepsilon^{k+2}) = \varepsilon^{k+1} \int_{\gamma_h} q_k\omega + O(\varepsilon^{k+2}),$$

i.e. $I_{k+1}(h) = \int_{\gamma_h} q_k\omega$. $\qquad \square$

Two important questions arise:

(i) Is there an integer K such that the above procedure stops at order K (i.e. if $I_j(h) \equiv 0$ for $j = 1, \ldots, K$, then all $I_j(h) \equiv 0$ for $j > K$)?

(ii) What kind of functions H satisfy the condition $(*)$?

The answer to the first question is positive, but a new problem is that there is no efficient method to find such integer K. Usually, the condition $I_j(h) \equiv 0$ for $j = 1, \ldots, l$ gives restrictions on the parameters which appear on the 1-form ω, and one needs to check if $dH - \varepsilon\,\omega$ is integrable at this stage. If the answer is yes, then $K = l$.

Concerning the second question, the following result by L. Gavrilov shows that for "generic" polynomial Hamiltonian H, the condition $(*)$ holds. Before stating the result, we need to give some definitions.

Definition 2.11 ([112]). A polynomial $f \in \mathbb{R}[x, y]$ is called *weighted homogeneous of weighted degree j and type (w_x, w_y)* if there are $w_x, w_y \in \mathbb{N}^+$ and $j \in \mathbb{N}$, such that

$$f(z^{w_x} x, z^{w_y} y) = z^j f(x, y), \quad \forall z \in \mathbb{R}.$$

A polynomial $f \in \mathbb{R}[x, y]$ is called *semiweighted homogeneous of weighted degree k and type (w_x, w_y)* if it can be written as $f = \sum_{j=0}^{k} f_j$, where f_j are weighted homogeneous polynomials of weighted degree j and type (w_x, w_y), and the polynomial $f_k(x, y)$ has an isolated critical point at the origin.

Theorem 2.12 (Proposition 3.2 of [112]). *Let $\gamma_h \subset H^{-1}(h) \subset \mathbb{R}^2$ be a continuous family of ovals surrounding a single critical point of H. If $H \in \mathbb{R}[x, y]$ is a semiweighted homogeneous Morse polynomial with distinct critical values, then the space of all real polynomial 1-forms satisfies the condition $(*)$.*

The above result was obtained by Y. Ilyashenko [164] when H is a polynomial of degree m with $(m-1)^2$ distinct critical points. Some further discussions concerning Theorem 2.10 can be found in [117].

Example 2.13. It was shown in [112] that if $P_m(x)$ is a polynomial of degree m with $m-1$ distinct critical values, then the Hamiltonian function $H(x, y) = \frac{1}{2}y^2 + P_m(x)$ satisfies the condition $(*)$. This result was found earlier in [90] for the case $H = \frac{1}{2}(x^2 + y^2)$. $\square$

The result in Theorem 2.10 was generalized by I.D. Iliev in [158] to the case $\omega = \omega_0 + \varepsilon\omega_1 + \varepsilon^2\omega_2 + \cdots$. He considers polynomial perturbations of Hamiltonian systems with elliptic or hyperelliptic Hamiltonians and gives a formula for the second variation of the displacement function in terms of the coefficients of the perturbations. We briefly introduce this result below.

Let $H(x, y) = \frac{y^2}{2} - U(x)$, where $U(x)$ is a polynomial of degree $n \geq 2$.

Consider the perturbations

$$\dot{x} = \frac{\partial H(x,y)}{\partial y} + \varepsilon\, f(x,y,\varepsilon),$$

$$\dot{y} = -\frac{\partial H(x,y)}{\partial x} + \varepsilon\, g(x,y,\varepsilon),$$

(2.7)

where $f(x,y,\varepsilon)$ and $g(x,y,\varepsilon)$ are polynomials in x,y which depend analytically on the small parameter ε. System (2.7) can be written in its Pfaffian form

$$dH - \varepsilon\,\omega = 0, \quad \omega = \omega_0 + \varepsilon\omega_1 + \varepsilon^2\omega_2 + \cdots,$$

(2.8)

where $\omega_0 = -f(x,y,0)dy + g(x,y,0)dx$, $\omega_1 = -f_\varepsilon(x,y,0)dy + g_\varepsilon(x,y,0)dx, \ldots$, and the subscript ε denotes the derivative with respect to ε.

Suppose that the continuous family of ovals $\gamma_h \subset H^{-1}(h)$ for $h \in (a,b)$, which form an annulus D. If we take a transversal segment to $\{\gamma_h\}$ and parameterize it using the level value h, then, by Theorem 2.1, the displacement function for small ε has the form

$$d(h,\varepsilon) = \varepsilon\, M_1(h) + \varepsilon^2\, M_2(h) + \cdots,$$

where $M_1(h) = \oint_{\gamma_h} \omega_0$.

Theorem 2.14 ([158]). *Under the above assumptions the following statements hold.*

(i) *If $M_1(h) \equiv 0$, then there exist in D a continuous function $q_0(x,y)$ and a locally Lipschitz continuous function $Q_0(x,y)$ such that the form ω_0 can be expressed as $\omega_0 = q_0\, dH + dQ_0$, and*

$$M_2(h) = \oint_{\gamma_h} (q_0\omega_0 + \omega_1).$$

(ii) *If $M_k(h) \equiv 0, 1 \le k \le m$, then define the 1-forms $\Omega_0, \Omega_1, \ldots, \Omega_m$ successively as follows:*

$$\Omega_0 = \omega_0, \quad \Omega_k = \omega_k + \sum_{i+j=k-1} q_i\,\omega_j, \quad \text{for } 1 \le k \le m,$$

where, for $0 \le k \le m-1$, the 1-form Ω_k can be expressed as $\Omega_k = q_k\, dH + dQ_k$ with q_k, Q_k as in statement (i), and

$$M_{m+1}(h) = \oint_{\gamma_h} \Omega_m.$$

(iii) *The function $M_2(h)$, $h \in (a, b)$ can be explicitly expressed as*

$$M_2(h) = \oint_{\gamma_h} [G_{1h}(x, y)P_2(x, h) - G_1(x, y)P_{2h}(x, h)]dx$$
$$- \oint_{\gamma_h} \frac{F(x, y)}{y}[f_x(x, y, 0) + g_y(x, y, 0)]dx \qquad (2.9)$$
$$+ \oint_{\gamma_h} g_\varepsilon(x, y, 0)dx - f_\varepsilon(x, y, 0)dy,$$

where

$$F(x, y) = \int_0^y f(x, s, 0)ds - \int_0^x g(s, 0, 0)\, ds,$$
$$G(x, y) = g(x, y, 0) + F_x(x, y),$$

and $G_1(x, y)$ and $G_2(x, y)$ are the odd and even parts of $G(x, y)$ with respect to y. That is, $G(x, y) = G_1(x, y) + G_2(x, y)$, $G_1(x, y) = y\, p_1(x, y^2)$, and $G_2(x, y) = p_2(x, y^2)$. Finally, $P_2(x, h)$ is the polynomial

$$P_2(x, h) = \int_0^x p_2(s, 2h + 2U(s))\, ds.$$

We notice that, in the last statement of the above result, the subscripts x, y, h, ε denote the first derivative with respect to x, y, h, ε, respectively. Moreover, $G_{1h}(x, y) = G_{1y}(x, y)/y$ on each oval $\gamma(h)$. If the divergence $f_x + g_y$ is either an odd or an even function of y, then formula (2.9) can be written in the more compact form:

$$M_2(h) = -\oint_{\gamma_h} \frac{F(x, y)}{y}[f_x + g_y]|_{\varepsilon=0}\, dx + \oint_{\gamma_h} (g_\varepsilon\, dx - f_\varepsilon\, dy)|_{\varepsilon=0}.$$

Remark 2.15. In 1995, before the general result in Theorem 2.10, B. Li & Z. Zhang [176] deduced the second-order Melnikov function $M_2(h)$ for the Bogdanov–Takens bifurcation problem (in codimension 2 case), see also [374]. In Example 2.16 we introduce this result, together with a result of [160] on $M_k(h)$ for arbitrary k. Note that the result in [176] not only provides the upper bound of the number of zeros of the second-order Melnikov function, but also it extends them to be the maximal number of limit cycles by using the technique from [253] (if $M_2(h)$ does not vanish identically). [160] also gives this extension of the result for $k \geq 3$. In Examples 2.18 and 2.19 we introduce applications of Theorem 2.10 to the general problem of degree 3 Hamiltonian perturbed also with degree 3 polynomials.

Example 2.16. For the Hamiltonian (1.6) under polynomial perturbations of arbitrary degree n, the following statements hold:

(i) If $M_1(h) \equiv 0$, then $M_2(h)$ has at most $2(n-1)$ zeros for n even and at most $2n - 3$ zeros for n odd, counting the multiplicities. See [176].

(ii) If $M_k(h)$ is the first Melnikov function in (2.6) which does not vanish identically, then $M_k(h)$ has no more than $k(n-1)$ zeros, counting the multiplicities. Moreover, if $k \geq 3$ (or $k \geq 2$ when n is odd) then $M_k(h)$ has no more than $k(n-1)-1$ zeros, counting the multiplicities. This number is sharp for $k = 3$. See [160]. $\qquad\square$

We notice that, in Example 2.16 above, the number of parameters in (2.8) also increase with k, but the above upper bound, for the number of zeros of M_k, do not take into account that the previous vanish identically. Clearly in each step the function M_k also includes the previous Melnikov functions. As it is also commented in [160] this estimate, in general, is only sharp for the first k. Usually, the number of zeros of M_k, which maintains or increases with k, stabilizes, but the explicit value for this case and its dependence on n is unknown. In next Example 2.17 is shown that the stabilization process, for classical Liénard family, occurs at the first step. That is all the Melnikov functions are equivalent. In the following Examples 2.18 and 2.19 the situation is very different because the perturbation (2.8) has only one term in ε, the corresponding to ω_0. That is the perturbation polynomials f and g in (2.7) do not depend on ε. This explains the conclusions of the first statements.

Example 2.17 ([109]). Let p_n be a polynomial of degree n such that $p_n(0) = 0$ and $p_n(-x) = p_n(x)$ for all x. If $M_j(h) \equiv 0$ for $j = 1, \ldots, k - 1$, the number of zeros of $M_k(h)$ for

$$\dot{x} = -y + p_n(x) + \varepsilon\, f_n(x, \varepsilon),$$

$$\dot{y} = x,$$

is $[(n-1)/2]$ for any k, when f_n is a polynomial of degree n such that $f(0, \varepsilon) = 0$. Clearly, the unperturbed system, $\varepsilon = 0$, is a reversible center. $\qquad\square$

The proof of the above example is based on the fact that, for every k, the first non-vanishing Melnikov function is always a polynomial with only odd monomials. In particular, at each step k, the conditions $M_k(h) \equiv 0$ force that the perturbed system also has a center and it can be integrated inside p_n. Hence, all functions M_k have the same "shape". Consequently, the number of zeros stabilizes at the first step.

Example 2.18 ([118]). For the Hamiltonian (1.13), excluding from consideration all symmetric cases, under polynomial perturbations of degree 3 the following statements hold:

(i) If $M_1(h) = M_2(h) = M_3(h) = M_4(h) \equiv 0$, then the perturbed system is integrable. That is $M_k(h) \equiv 0$ for $k \geq 5$.

(ii) The order k Melnikov functions $M_k(h) = \alpha_k(h)I_0(h) + \beta_k(h)I_1(h) + \gamma_k(h)I_2(h)$, where $1 \leq k \leq 4$, $\alpha_k(h), \beta_k(h)$, and $\gamma_k(h)$ are polynomials of degree at most one, being $I_m(h) = \oint_{\gamma_h} x^m y\, dx$, for $m = 0, 1, 2$. Moreover, $M_2(h)$ has the maximum possible number of zeros among $M_k(h)$.

(iii) In the interior eight-loop case, at most five limit cycles bifurcate from each one of the annuli inside the loop; in the exterior eight-loop case, at most eight limit cycles bifurcate from the annulus outside the loop; in the saddle-loop case, at most seven limit cycles bifurcate from the unique period annulus. See Figure 3 for the phase portraits of X_H.   □

Example 2.19 ([162]). If the Hamiltonian (1.13) exhibits a symmetric eight-loop, we can transform it into the form $H = (2y^2 - 2x^2 + x^4)/4$, X_H has a saddle at the origin and two centers at $(1,0)$ and $(-1,0)$. When $h \in (-1/4,0)$ there are two symmetric continuous families of ovals $\{\gamma_h\} \subset H^{-1}(h)$. Under polynomial perturbations of degree three the following statements hold:

(i) If $M_1(h) = M_2(h) = M_3(h) = M_4(h) \equiv 0$, then the perturbed system is integrable.

(ii) $M_k(h) = \alpha_k(h)I_0(h) + \beta_k(h)I_1(h) + \gamma_k(h)I_2(h)$, where $1 \leq k \leq 4$, $\alpha_k(h)$, $\beta_k(h)$, and $\gamma_k(h)$ are polynomials of degree at most one, $I_m(h) = \oint_{\gamma_h} x^m y\, dx$, for $m = 0,1,2$, and $I_1(h) = \sqrt{2}\pi\,(h + 1/4)$. Moreover, $M_2(h)$ has the maximum possible number of zeros among $M_k(h)$.

(iii) At most five limit cycles bifurcate from each one of the annuli inside the loop, and this bound is exact; at most nine limit cycles simultaneously bifurcate from the annuli.   □

Remark 2.20. When the condition $(*)$ is not satisfied, it is also possible to use the recursion formula of Françoise (Theorem 2.10), but the functions q_k and R_k may not be polynomials, see [157] for example, where q_k and R_k have rational functions besides some logarithms. These logarithms appear naturally when we consider the perturbation of the linear center, $H = (x^2 + y^2)/2$, having a straight line of singularities. That is, when instead of (2.8) we consider $(1+x)dH - \varepsilon\omega = 0$. Up to a third-order analysis of this problem is studied in [20]. The logarithms appear also in [284] when first and second-order analysis are analyzed, but in polar coordinates, for quadratic systems having a rational first integral. Of course, in this case the study of "generalized Abelian integrals" is more difficult, see the next section.

Remark 2.21. Using the Françoise's procedure, A. Gasull & J. Torregrosa constructed a new algorithm of the computation of the usual Lyapunov constants in [108]. With a generalized algorithm, they study the generalized Lyapunov constants for some degenerate singularities in [107]. Additionally, in [109], the relation between the degenerate Hopf bifurcation and the method of Abelian integrals near the singularity is analyzed. More concretely, when $H = (x^2 + y^2)/2$ and the 1-forms ω_k in (2.8) are homogeneous polynomials, both bifurcation mechanisms provide the same information.

Remark 2.22. In the paper [328] M. Viano, J. Llibre & H. Giacomini gave a different recursive procedure for the calculation of the higher-order Melnikov functions using the reciprocal integrating factor.

2.3 The Integrable and Non-Hamiltonian Case

For the complete study of the Hilbert's 16th problem, we need to consider the cyclicity of a period annulus (or annuli) under polynomial perturbations not only from the polynomial Hamiltonian systems, as we have explained in the last two sections, but also from polynomial integrable and non-Hamiltonian systems. To see it clearly, let us list all integrable quadratic systems with at least one center. By using the terminology from [393], Iliev [159] classified them into the following five classes using complex notation:

(1) $\dot{z} = -\mathrm{i}\,z - z^2 + 2z\bar{z} + (b + \mathrm{i}\,c)\bar{z}^2$, Hamiltonian ($Q_3^H$),

(2) $\dot{z} = -\mathrm{i}\,z + az^2 + 2z\bar{z} + b\bar{z}^2$, reversible ($Q_3^R$),

(3) $\dot{z} = -\mathrm{i}\,z + 4z^2 + 2z\bar{z} + (b + \mathrm{i}\,c)\bar{z}^2$, $|b + \mathrm{i}\,c| = 2$, codimension 4 (Q_4),

(4) $\dot{z} = -\mathrm{i}\,z + z^2 + (b + \mathrm{i}\,c)\bar{z}^2$, generalized Lotka–Volterra (Q_3^{LV}),

(5) $\dot{z} = -\mathrm{i}\,z + \bar{z}^2$, Hamiltonian triangle,

where the parameters a, b, and c are real, and $z = x + \mathrm{i}\,y$. Then, an integrable quadratic system is called *generic*, if it belongs to one of the first four classes and does not belong to other classes of the classification given above. Otherwise, it is called *degenerate*.

The discussions about Abelian integrals so far were only valid for the generic Hamiltonian class. As an example of integrable and non-Hamiltonian case, we consider the reversible class. Taking $z = x + \mathrm{i}\,y$, and the change $t \mapsto -t$, we obtain that the above family (2) writes as

$$\dot{x} = -y - (a + b + 2)x^2 + (a + b - 2)y^2,$$
$$\dot{y} = x - 2(a - b)xy.$$

If $c = a - b \neq 0$, then making the scaling $(x, y) \mapsto (x/c, y/c)$ and changing the parameters $\left(-\frac{a+b+2}{a-b}, \frac{a+b-2}{a-b}\right) \mapsto (\alpha, \beta)$, we obtain

$$\dot{x} = -y + \alpha x^2 + \beta y^2,$$
$$\dot{y} = x(1 - 2y). \tag{2.10}$$

Using the following coordinates and time scaling,

$$x = \frac{1}{2}\hat{x}, \quad y = -\frac{1}{2}(\hat{y} - 1), \quad t = 2\hat{t},$$

and then writing again, for simplicity, (x, y, t) instead of $(\hat{x}, \hat{y}, \hat{t})$, we obtain

$$\dot{x} = \alpha x^2 + \beta y^2 - 2(\beta - 1)y + (\beta - 2),$$
$$\dot{y} = -2xy. \tag{2.11}$$

System (2.11) has the invariant straight line $\{y = 0\}$ and a center at $(0, 1)$. The singularity $(0, (\beta - 2)/\beta)$ is also a center if $0 < \beta < 2$, and is a saddle if $\beta < 0$ or $\beta > 2$. If $\alpha(2 - \beta) > 0$, then the system has two saddles at $(\pm\sqrt{(2 - \beta)/\alpha}, 0)$.

If $\alpha(\alpha + 1)(\alpha + 2) \neq 0$, then the first integral of system (2.11) is given by

$$F(x, y) = |y|^\alpha (x^2 + Ly^2 + My + N) = h, \tag{2.12}$$

where

$$L = \frac{\beta}{\alpha + 2}, \quad M = \frac{2(1 - \beta)}{\alpha + 1}, \quad N = \frac{\beta - 2}{\alpha}.$$

Note that if $\alpha \neq 1$, then system (2.11) is not Hamiltonian, and the integrating factor is $\mu = |y|^{\alpha - 1}$. In the period annulus surrounding a center of system (2.11), we denote the ovals by

$$\Gamma_h = \{(x, y) \in \mathbb{R}^2 : F(x, y) = h, \ h_c < h < h_s\},$$

where h_c is the critical value of H at a center, and h_s is the value of H for which the period annulus ends at a separatrix polycycle or at infinity. We can suppose that $h_c < h_s$, otherwise we can change the sign of H to ensure it.

We consider quadratic perturbations of (2.11):

$$\dot{x} = \alpha x^2 + \beta y^2 - 2(\beta - 1)y + (\beta - 2) + \varepsilon f(x, y),$$
$$\dot{y} = -2xy + \varepsilon g(x, y),$$

where ε is a small parameter, and the perturbations functions f and g are quadratic polynomials in x and y.

If $\alpha \neq 1$, $\alpha + \beta \neq 0$, $(\alpha, \beta) \neq (-4, 2)$, and $(\alpha, \beta) \neq (-2/3, 0)$, then the reversible system (2.11) is generic. If, in addition, $\alpha(\alpha + 1)(\alpha + 2) \neq 0$, then the cyclicity of the period annulus of (2.11) under quadratic perturbations is equal to the maximal number of isolated zeros in (h_c, h_s), counting multiplicities, of the following integral (see [159])

$$M_1(h) = \int_{\Gamma_h} |y|^{\alpha - 2}(\hat{\alpha} + \hat{\beta}y + \hat{\gamma}y^2)\, x\, dy, \tag{2.13}$$

where $\hat{\alpha}$, $\hat{\beta}$, and $\hat{\gamma}$ are real constants, and the orientation of the integral is given by the vector field.

It is clear now that for most values of α, say α is not an integer, both the first integral F in (2.12) and the integrand function in (2.13) are no longer polynomials. Hence, the integral $M_1(h)$ is not an Abelian integral in the strict meaning. Usually it is called *generalized* Abelian integral or *pseudo-Abelian* integral, or simply Abelian integral as before.

In Theorem 1 of [110] S. Gautier, L. Gavrilov & I.D. Iliev classify all quadratic reversible systems, whose phase curves are algebraic curves of genus one, into 18 cases (r1)–(r18). They use the complex form of the integrable quadratic systems Q_3^R with a center at the origin as follows

$$\dot{z} = -\mathrm{i}\, z + a z^2 + 2 z \bar{z} + b \bar{z}^2, \quad a, b \in \mathbb{R}, \quad z = x + \mathrm{i}\, y, \qquad (2.14)$$

and give the following conjecture.

Conjecture 2.23 ([110]). *The period annulus around the center at the origin in (r1)–(r18) has the following cyclicity under small quadratic perturbations: three for cases (r1) with $a^* < a < 4$, $b = (a-1)/2$, (r3) with $7/3 < a < 4$, $b = (a-4)/5$, (r4) with $4 < a < 5$, $b = -(a+4)/3$, (r5) with $a = 4$, $b = 2$, (r6) with $a > 4$, $b = 3a + 2$, and (r10), and two otherwise.*

Note that $a^* \in (5/3, 3)$ is determined from a transcendental equation (see [161]) and it can be calculated numerically, $a^* \approx 2.0655$. Moreover, in the above conjecture, other values of (a, b) that are equivalent due to symmetry transformations are not detailed. For example, following [161], the values $4 < a < \hat{a}^*$ for $\hat{a}^* \approx 4.7539$ are equivalent to the described ones.

If we transform system (2.14) into system (2.10), as we did above (note that for genus one cases we have $a - b \neq 0$), then the conditions (r1)–(r18) can be rewritten in terms of the parameters α and β, (see [180]). Several authors verify the above conjecture for different cases, we briefly describe their results below together with the new notation.

(r1) This family corresponds to $\alpha = -3$ and it was completely studied in different works. See [82] for $\beta = 1$ (also see [252] for a simple proof about the number of zeros in one annulus by using the Chebyshev property); [263] for $\beta = -1$; [356] for $\beta \in (-\infty, 0) \setminus \{-1\}$; [161] for $\beta \in (0, 2)$; [162] for $\beta \in [2, +\infty) \setminus \{3\}$; [189] for $\beta = 3$ (a reversible and Lotka–Volterra case), and [306] for $\beta = 0$.

(r2) This family corresponds to $\alpha = 1$ and it is the intersection of reversible and Hamiltonian classes. The cyclicity problem of the period annulus or annuli was completely solved, see the introduction above ([54, 114, 156, 157, 385, 389], and [188]).

(r3) This family corresponds to $\alpha = -3/2$ and $\beta \neq 2$. It was also completely studied also in different papers. See [219] for $\beta \in (0, 2)$; [388] for $\beta \in (2, +\infty)$, and [217] for $\beta \in (-\infty, 0] \cup \{2\}$.

(r4) This family corresponds to $\alpha = -1/2$ and $\beta \neq 0$. Only the interval $\beta \in (0, 2)$ is studied. See [65].

(r5) This family corresponds to $\alpha = -4$ and a complete study has been almost finished. See Theorem 2.1 of [33] for $\beta \neq 0, 2, 4$.

(r6) This family corresponds to the value $\alpha = 2$. It was studied in Theorem 1.1 of [33] for β in the interval $(0, 2)$.

(r7) This system was studied in [129, 268] and corresponds to the pair $(\alpha, \beta) = \left(-\frac{4}{3}, 0\right)$.

(r8) This system was studied in [39]. It corresponds to the pair $(\alpha, \beta) = (-\frac{2}{3}, 2)$.

(r9) This system was studied in [271]. It corresponds to the pair $(\alpha, \beta) = (-\frac{4}{3}, 2)$.

(r10) This system was studied in [159]. It corresponds to the pair $(\alpha, \beta) = (-\frac{2}{3}, 0)$.

(r11) This system was studied in [110, 129]. It corresponds to the pair $(\alpha, \beta) = (-\frac{5}{3}, 2)$.

(r12) This system was studied in [268]. It corresponds to the pair $(\alpha, \beta) = (-\frac{1}{3}, 0)$.

(r13) This system was studied in [39]. It corresponds to the pair $(\alpha, \beta) = (-\frac{5}{4}, 2)$.

(r14) This system was studied in [129, 341]. It corresponds to the pair $(\alpha, \beta) = (-\frac{3}{4}, 0)$.

(r15) This system was studied in [129, 264]. It corresponds to the pair $(\alpha, \beta) = (-\frac{7}{4}, 2)$.

(r16) This system was studied in [39, 270]. It corresponds to the pair $(\alpha, \beta) = (-\frac{1}{4}, 0)$.

(r17) This system was studied in [129]. It corresponds to the pair $(\alpha, \beta) = (-\frac{5}{2}, 2)$.

(r18) This system was studied in [110]. It corresponds to the pair $(\alpha, \beta) = (\frac{1}{2}, 0)$.

From the above list, it is clear that each case from (r1) to (r6) corresponds to a one-parameter family of systems, and each one of the other corresponds to a single system with a value $\beta = 0$ or $\beta = 2$ and some special value of α.

Note that if $\beta \in (0, 2)$ in cases (r1)–(r6), then system $X_{\alpha,\beta}$ has two centers and two period annuli. The relevant papers studied the maximal number of zeros of Abelian integral for each annulus and also for two annuli at the same time. In fact, the bifurcation diagrams in the parameter space were obtained. Besides, [221] proved that the cyclicity under quadratic perturbations is two when $\beta = \alpha + 2$, $\alpha(\alpha + 2) > 0$, and $\alpha \neq 1$. A quadratic system possessing at least 4 limit cycles (in (3,1)-distribution), constructed from a quadratic reversible system by quadratic perturbations, was found in 1982 in [177]. If $a = 0$ then system (30) of this last referred paper is reversible, see the first footnote on page 1093 of it. In 2012, [220] studies this problem in family (r3) ($\beta \in (0, 2)$) and [359] studies this problem by using different normal forms.

As we have commented above, family (r1) has two centers when $\beta \in (0, 2)$. In [161], 4 limit cycles near family (r1), in (3, 1)-distribution, are obtained studying the simultaneity of zeros of three Abelian integrals, as the ones defined in (2.13), in the two period annuli. As we have only three Abelian integrals the Chebyshev property given in Definition 1.14 can not be satisfied. More concretely, the Wronksian of the three Abelian integrals, defined in Proposition 1.15, vanishes at most once in each interval of definition, depending on the value of the parameter β. We will give more details on this fact at the end of Section 3.2.

A simple study of the local cyclicity of the origin of the reversible center (2.10) in terms of the parameters (α, β) can be done using the return map near the origin of the perturbed system

$$\dot{x} = -y + \alpha x^2 + \beta y^2 + a_{20}x^2 + a_{11}xy + a_{02}y^2,$$
$$\dot{y} = x(1 - 2y) + b_{20}x^2 + b_{11}xy + b_{02}y^2.$$

Using the usual complex notation, $\dot{z} = iz + Az^2 + Bz\bar{z} + C\bar{z}^2$, the well-known Lyapunov quantities (see [393]) for a quadratic system write as

$$L_1 = -2\pi \operatorname{Im}(AB),$$
$$L_2 = -\frac{2\pi}{3}(2A + \bar{B})(A - 2\bar{B})\bar{B}C,$$
$$L_3 = -\frac{5\pi}{4}(|B|^2 - |C|^2)(2A + \bar{B})\bar{B}^2C.$$

So, the first-order Taylor series with respect to the perturbation parameters $(a_{20}, a_{11}, a_{02}, b_{20}, b_{11}, b_{02})$ allow us to write them as

$$\hat{L}_1 = \frac{\pi}{2}\big((\alpha+\beta)a_{11}+2(\beta+1)b_{02}-2(\alpha-1)b_{20}\big),$$
$$\hat{L}_2 = \frac{\pi}{48}\big((\alpha+\beta)(5\alpha^2+6\alpha\beta+5\beta^2-8\alpha-16\beta-12)a_{11}$$
$$- 2(12\alpha^3+15\alpha^2\beta-6\alpha\beta^2-5\beta^3+15\alpha^2+22\alpha\beta+11\beta^2-8\alpha-12\beta-4)b_{02}$$
$$- 2(\alpha-1)(5\alpha^2-6\alpha\beta-15\beta^2+16\alpha+16\beta+4)b_{20}\big),$$
$$\hat{L}_3 = -\frac{5\pi}{1024}(\alpha+3\beta-2)(3\alpha+\beta+2)\big((\alpha+\beta-4)(\alpha+\beta)^2a_{11}$$
$$+2(\alpha+\beta)(4\alpha^2-\alpha\beta-\beta^2+3\alpha+3\beta-4)b_{02}+2(\alpha-1)(\alpha+\beta)(\alpha-3\beta+4)b_{20}\big).$$

As $\alpha \neq 1$, using the Implicit Function Theorem, there exists a change of variables in the parameter space that writes $L_1 = \hat{L}_1 = u_1$. Hence, under the condition $L_1 = u_1 = 0$ we have

$$\hat{L}_2 = -\frac{\pi}{6}(\alpha-1)(3\alpha+5\beta+2)\big((\alpha+2)b_{02} + (2-\beta)b_{20}\big),$$
$$\hat{L}_3 = \frac{5\pi}{128}(\alpha-1)(\alpha+3\beta-2)(\alpha+\beta)(3\alpha+\beta+2)\big((\alpha+2)b_{02} + (2-\beta)b_{20}\big).$$

Generically, that is when $3\alpha + 5\beta + 2 \neq 0$ because $\alpha \neq 1$, there exists a local change of coordinates such that $L_2 = u_2$. Under this generic condition, when $L_2 = u_2 = 0$ also $L_3 = 0$ and, consequently, the local cyclicity is only 2. Here we have also used the trace parameter to have a versal unfolding of the weak focus of order two. This property is used in [393] to justify that the reversible family is of codimension three and is labeled as Q_3^R. But it remains a carefully study on the straight line $3\alpha + 5\beta + 2 = 0$. Because an extra limit cycle appears. In fact, on this straight line, as $\hat{L}_3 \neq 0$ when $\beta \notin \{0, -1, 2\}$, we have a weak focus of order three that unfolds three limit cycles of small amplitude, because it is simple on the decomposition of $\hat{L}_2$. For a proof of this general result see [121]. We notice that the value of (α, β) for system (r10) is on this special line. Moreover, this provides new values where the Conjecture 2.23 is satisfied. Finally, we remark that this is the same bifurcation phenomenon that explains the existence of 12 limit cycles bifurcating from the origin in the cubic family, see [362] or again [121], for a more detailed proof as we have explained in Chapter 1.

The perturbations with polynomials up to degree n for cases (r9)–(r18) were studied by X. Hong and his collaborators trying to find some upper bounds for the corresponding Abelian integrals, see [148, 149, 150]. In these papers also the perturbation of four more families were considered. The ones labeled as (r19)–(r22) in [110], which are the ones that the ovals are ellipses. They correspond, in our notation, to the pairs $(\alpha, \beta) \in \{(-3/2, 2), (-1/2, 0), (-2, 0), (0, 2)\}$. Systems (r20) and (r21) are the isochronous centers S_3 and S_2, respectively. The isochronous center S_4 is in family (r5) and it corresponds to $(\alpha, \beta) = (-4, 1)$. The last quadratic isochronous center S_1, which is $(\alpha, \beta) = (-1, 1)$ has not any correspondence with the original parameters (a, b). We recall that the perturbation of isochronous centers is analyzed in [51] and the isochronous center classification is due to W.S. Loud [249]. See also [31].

Concerning the quadratic perturbations of period annuli in quadratic reversible and Lotka–Volterra systems $(Q_3^R \cap Q_3^{LV}$, i.e. $\alpha + \beta = 0$ in (2.10)), several results give cyclicity 3 or 2. If the phase curves are algebraic curves of genus one, then [110] classifies them into 6 cases (rlv1)–(rlv6), and also proposed a conjecture about the cyclicity. Above mentioned [156] and [190] studied the cases (rlv1) and (rlv2), respectively, and [129], [305] and [267] studied the cases (rlv3), (rlv4), and (rlv6), respectively. Besides, [189] studied the case $(\alpha, \beta) = (-1, 1)$, which is the above mentioned quadratic isochronous center S_1, by using second-order averaging. Recently, the perturbation of the isochronous quadratic centers S_1, S_2, and S_3 has been revisited in [284], using the special property that they are linearizable via a birational transformation. For more details see Section 2.4.

Using Abelian integrals, based on Picard–Fuchs equations and argument principle, L. Gavrilov & I.D. Iliev [115] proved that the cyclicity of the period annulus of quadratic codimension four center (Q_4) under quadratic perturbations is less than or equal to eight, then Y. Zhao improved this number decreasing the upper bound to five, see [381].

Finally, let us give a general setting for the integrable and non-Hamiltonian case. Suppose that the unperturbed system has a first integral $F(x, y)$ with an integrating factor $\mu(x, y) = 1/R(x, y)$; then the perturbed system can be written in the form

$$\begin{cases} \dot{x} = -\dfrac{\partial F(x, y)}{\partial y} R(x, y) + \varepsilon f(x, y), \\[2mm] \dot{y} = \dfrac{\partial F(x, y)}{\partial x} R(x, y) + \varepsilon g(x, y), \end{cases}$$

and, associated to it, we define the (generalized) Abelian integral

$$I(h) = \int_{\gamma_h} \frac{f(x,y)\,dy - g(x,y)\,dx}{R(x,y)},$$

where $\{\gamma_h\}$ are the family of ovals contained in the level curves $\{F(x,y) = h\}$. By the same mechanisms as in the last sections, the integral $I(h)$ gives the first approximation of the displacement function.

Note that some traditional methods, such as the derivation of the Picard–Fuchs equation or the Picard–Lefschetz formula, among others, fail for this generalized form of the Abelian integrals. For this reason, the analysis of the perturbations from the reversible class Q_3^R is very difficult. On the other hand, this study, comparing with the quadratic perturbations from other quadratic integrable classes, is the most interesting one. We list some results concerning the quadratic perturbations from the reversible system (2.10): [50] for the isochronous centers; [316] for the bifurcation curve of the unbounded heteroclinic loop; [82, 161, 263, 356] for $a = -3$ with different values of b; [33] for $a \sim 2$ with different values of b. There are some other works dealing with the perturbations from integrable and non-Hamiltonian systems, among them [8, 64, 103, 185, 186, 215, 239, 384].

2.4 The Study of the Period Function

We use the same notation as before to denote a continuous family of ovals $\gamma_h \subset H^{-1}(h)$, where H is the Hamiltonian function (or the first integral) of a planar Hamiltonian (or integrable) system. Each γ_h is a periodic orbit of the system, so we have a *period function* $T(h)$, parameterized by the same $h \in (a, b)$. If the period function is constant for all h, then the period annulus is *isochronous*. If the isochronous period annulus surrounds a center, then the center is called isochronous. If the period function is strictly increasing or decreasing, then we say that the period function is *monotone*. Otherwise, the period function has *critical points*.

The study of the period function and the study of the Abelian integral have some relations, at least from the following two points of view.

First, the study of the number of critical points of the period function by perturbing an isochronous center inside a certain class of integrable systems, is comparable to the study of the number of zeros of an Abelian integral by perturbing an integrable system inside a certain class of systems. Later we will introduce a work by E. Freire, A. Gasull & A. Guillamon [93] on this aspect.

Second, the study of the period function is useful for the study of the Abelian integral. For example, the Abelian integral $I_0(h) = \oint_{\gamma_h} y\,dx$ gives the area of the region surrounded by γ_h. Here we suppose that the orientation of γ_h is clockwise, $\gamma_h \subset H^{-1}(h)$, and the related Hamiltonian system is $X_H = H_y \partial/\partial x - H_x \partial/\partial y$. As above, the subscripts x and y denote the respective first derivatives. Hence,

$I_0(h) > 0$ for $h > a$ and the derivative $I_0'(h) > 0$ gives the period of γ_h. In fact,

$$I_0'(h) = \oint_{\gamma_h} \frac{\partial y}{\partial h}\, dx = \oint_{\gamma_h} (H_y)^{-1}\, dx = \oint_{\gamma_h} dt = T(h).$$

If the period function is monotone, then $I_0''(h) \neq 0$. In some studies of Abelian integrals this information is needed to define a function by a ratio of two Abelian integrals with second-order derivative, see for example [32, 33, 54, 78, 79, 81, 161] and Section 4.2 below. If $I_0''(h)$ has some zeros, i.e. the period function has critical points, then the use of the ratio becomes complicated, see [80]. For this reason, we will also briefly introduce some results on the period function.

Before stating the result of [93], we give the characterizations of isochronous centers: A center point p of a planar smooth vector field X in an annulus D is an isochronous center if and only if one of the following assertions holds:

(i) There exists a smooth change of coordinates in a neighborhood of p that linearizes X (a classical result of Poincaré).

(ii) There exists a transversal vector field U, commuting with X, i.e. $[X, U] = DU\,X - DX\,U = 0$, see [297, 334].

(iii) There exists a transversal vector field U and a scalar function μ such that $[X, U] = \mu X$ and
$$\int_0^{T_r} \mu(x(t), y(t))\, dt = 0,$$
where $\gamma = \{\varphi(t) = (x(t), y(t)), t \in [0, T_r]\}$ is any periodic orbit of X in D, and T_r is its period. In this case, U is called a *normalizer* of X, see [94].

Theorem 2.24 ([93]). *Suppose that a vector field X has an isochronous center of period T_0 in D. Consider a vector field U transversal to X such that $[X, U] = 0$. Let $\gamma(t) := \{\varphi(t; \psi(h)), t \in [0, T_0]\}$ be the set of periodic orbits of X in D parameterized by the time flow of U. Consider the family of vector fields $X_\varepsilon = X + \varepsilon Y$ having also a center; write Y as $Y = aX + bU$ and denote by $\gamma_\varepsilon(h)$ a generic closed orbit of X_ε passing through $\psi(h)$. The following statements hold:*

(i) *The period function associated to $\gamma_\varepsilon(h)$ is*
$$T_{\gamma_\varepsilon}(h) = T_0 + \varepsilon T_1(h) + O(\varepsilon^2),$$

where
$$T_1(h) = -\int_0^{T_0} a(\varphi(t; \psi(h)))\, dt.$$

(ii) *The derivative of T_1 with respect to h is:*
$$T_1'(h) = -\int_0^{T_0} \nabla a(x) \cdot U(x)\big|_{\{x = \varphi(t; \psi(h))\}}\, dt.$$

(iii) *If h^* is a simple zero of $T_1'(h)$, then for small ε there is exactly one critical period of X_ε close to h^* which tends to h^* as $\varepsilon \to 0$.*

We notice that the last statement of the above result is the equivalent of Theorem 2.4.(ii) for limit cycles, showing the parallelism between both similar problems. It is natural that in the study of the number of oscillations of the period function one can use the same scheme and techniques as in the study of the number of limit cycles. The next example proposes an Abelian integral for studying the oscillations of the period function inside the class of potential centers.

Example 2.25 ([93]). Consider the system

$$\dot{x} = -y,$$
$$\dot{y} = x + \varepsilon\, G'(x). \tag{2.15}$$

Then, for ε sufficiently small, the zeros of

$$I(s) = \int_0^{2\pi} \left. \frac{x(xG''(x) - G'(x))}{x^2 + y^2} \right|_{x=s\cos t,\, y=s\sin t} dt$$

give rise to critical periods of (2.15).

Moreover, if $G'(x)$ is a polynomial of degree n vanishing at zero, then the maximum number of simple zeros of $I(s)$ is $[(n-3)/2]$, i.e. at most $[(n-3)/2]$ critical periods bifurcate from the closed orbits of $(2.15)_{\varepsilon=0}$ in any fixed compact set in the period annulus region.

The above result can be proved by Theorem 2.24. In fact, let $X = (-y, x)$, $U = (x, y)$, and $Y = (0, G'(x))$. Then

$$a = \frac{x\, G'(x)}{x^2 + y^2}, \qquad b = \frac{y\, G'(x)}{x^2 + y^2},$$

and

$$\nabla a \cdot U = \frac{x(xG''(x) - G'(x))}{x^2 + y^2}.$$

Taking $\psi(h) = (e^h, 0)$, $\varphi(t; \psi(h)) = (e^h \cos t, e^h \sin t)$, and renaming e^h by s, the result follows by applying Theorem 2.24. □

We finish this section listing some results concerning the period function.

- A survey article about isochronous centers can be found in [31]. In this work appears a clear description of different families of isochronous centers. With emphasis in the families having homogeneous nonlinearities. For almost all of them the linearizations are detailed as well as the corresponding transversal vector field U.

- Every center of a polynomial Hamiltonian system of degree 4 (that is, with its homogeneous part of degree 4 not identically zero) is non-isochronous ([172]).

- Suppose $H(x,y) = F(x) + G(y)$ and the origin is a non-degenerate center of X_H. More concretely, if $T(h)$ denotes the period of the periodic orbit contained in $H(x,y) = h$, then [60] solved the inverse problem of characterizing all systems with a given function $T(h)$, characterized the limiting behavior of T at infinity when the origin is a global center and applied this result to prove, among other results, that there are no nonlinear polynomial isochronous centers in this family. Recently, this class of separable variables Hamiltonian systems is studied in [333], providing sufficient conditions for the period function to be monotone.

- An analytic vector field has a finite number of critical periods in any compact region inside an annulus ([49]). In fact, it is proved that if the set of critical periods in a period annulus of an analytic vector field on a compact subset of the plane or on the sphere is infinite, then the period annulus is isochronous.

- For the elliptic Hamiltonian $H(x,y) = y^2/2 + P(x)$, where P is a polynomial, the period function of the corresponding system X_H is monotone if $\deg(P) = 3$ and has at most one critical point if $\deg(P) = 4$. In the latter case only the global center (see case (iii) in Figure 3) has a critical period ([55, 111]).

- If the Hamiltonian system with Hamiltonian $H(x,y) = y^2/2 + V(x)$, where V is a smooth function, has a non-degenerate relative minimum at $x = 0$, then the period function is monotone if $V/(V')^2$ is convex ([46]). This condition was generalized in [17, 60].

- It was proved in [104] that for each n even there is a polynomial system of degree n having a center such that the period function associated to its period annulus has at least $(n^2 + 6n - 16)/4$ critical periods. The result follows perturbing the rigid and reversible system $(r', \theta') = (-r^n \sin\theta, 1)$ in the class of polynomial reversible systems of degree n. There is a similar result when n is odd. The considered unperturbed system is also the rigid even one with degree $n - 1$ but perturbed in the odd reversible class of degree n. Of course the bound will be also of order $n^2/4$. In 2021, X. Cen proved in [29] a higher lower bound of order $n^2/2$. Very recently, in 2023, and for $n \geq 3$ P. De Maesschalck & J. Torregrosa have improved the lower bound up to $n^2 - 2$ (resp. $n^2 - 2n - 1$) for n odd (resp. even), see [72].

The following results show the behavior of the period function for quadratic integrable systems with period annulus (or annuli). The notation is introduced at the beginning of the previous section.

- At most two *local critical periods* bifurcate from quadratic centers. Here local means when the perturbation parameter ε tends to zero, the level curves with bifurcating critical periods shrinks to the center point ([50]).

- It was conjectured in [46] that all the centers encountered in the family of second-order differential equations $\ddot{x} = V(x, \dot{x})$, being V a quadratic polynomial, should have a monotone period function, and some cases were solved in that paper. The remaining cases were completely solved in [99]. Note that this equation can be written as the planar system $(\dot{x}, \dot{y}) = (y, -x+ax^2+bxy+cy^2)$. The period function for the quadratic Hamiltonian systems ($X \in Q_3^H$) and for quadratic codimension 4 systems ($X \in Q_4$) is monotone. See [68] and [381], respectively. It is also monotone ([290, 299, 335]) for the *classical* quadratic Lotka–Volterra systems. But in general for $X \in Q_3^{LV}$ the problem is still open. Some partial results about the monotonicity of the period function in this class, especially the monotone property of the period function near the saddle loop, were obtained in [330].

- The behavior of the period function for $X \in Q_3^R$ (the family of quadratic reversible systems) is more complicated. The first example in this family with non-monotone period function was given in [48]. Then C. Chicone conjectured that the reversible centers have at most two critical periods (see Math. Review 94h:58072). As it was shown in the previous section, Q_3^R is a family of two parameters. The papers [288, 382, 383] analyze some 1-parameter families inside Q_3^R (including the example of [48]) and in them it is proved that at most one critical period may happen. In [329] several two-dimensional regions in the parameter plane were determined for which the corresponding center has a monotone period function. In [255] the behavior of the period function for $X \in Q_3^R$ is determined near the saddle loop. Some regions in the parameter plane were determined, for which the corresponding system has one or two critical periods near the saddle loop, and the local bifurcation diagram was also given. In the paper [332] a generalization of Q_3^R class, called generalized Loud systems, is introduced and the number of critical periods is studied.

- If $X \in Q_3^R \cap Q_3^{LV}$ (the family of quadratic reversible and Lotka–Volterra systems), then the period function is monotone except for the case that the system has a pair of isochronous centers, see [206].

- Recently, there is a survey article on period functions containing more references, see [181].

Chapter 3

Estimate of the Number of Zeros of Abelian Integrals

In the study of the weak Hilbert's 16th problem by using Abelian integrals, it is crucial to estimate the number of zeros of them. In this chapter, we introduce several methods to study the number of zeros of the Abelian integrals. In particular for the one defined in (1.11), which is related to the codimension 2 Bogdanov–Takens bifurcation problem, as we explained in Subsection 1.2.2.

3.1 The Method Based on the Picard–Fuchs Equation

Recall that we consider the elliptic Hamiltonian of degree 3 in the form

$$H(x,y) = \frac{y^2}{2} - \frac{x^3}{3} + x, \tag{3.1}$$

with the continuous family of ovals

$$\{\gamma_h\} = \{(x,y) : H(x,y) = h, \ -2/3 \le h \le 2/3\}, \tag{3.2}$$

shown in Figure 2. We denote the corresponding Hamiltonian differential system by X_H. When $h \to -2/3^+$ the oval γ_h shrinks to the center of X_H at $(-1,0)$ while when $h \to 2/3^-$ the oval γ_h terminates at the homoclinic loop of the saddle at $(1,0)$. The perturbation of X_H has the form

$$\begin{aligned}
\dot{x} &= y, \\
\dot{y} &= -1 + x^2 + \varepsilon(\alpha + x)y,
\end{aligned} \tag{3.3}$$

where α is a constant and ε is a small parameter. The corresponding Abelian integral is

$$I(h) = \alpha I_0(h) + I_1(h) \quad \text{with } I_j(h) = \oint_{\gamma_h} x^j y \, dx \text{ for } j = 0, 1, 2, \ldots \tag{3.4}$$

C. Christopher et al., *Limit Cycles of Differential Equations*, Advanced Courses in Mathematics - CRM Barcelona, https://doi.org/10.1007/978-3-030-59656-9_13

Since the orientation of γ_h is clockwise by (3.3), it is easy to check by using the Green's formula that $I_0(h)$ is the area of the region surrounded by γ_h, hence $I_0(h) > 0$ for $h > -2/3$. Let $(\xi_h, 0)$ and $(\eta_h, 0)$ $(\xi_h < -1 < \eta_h < 1)$ be the two intersection points of γ_h with the x-axis, then by using (3.1) it is easy to find

$$I_j(h) = 2 \int_{\xi_h}^{\eta_h} x^j y(x, h)\, dx, \quad I_j'(h) = 2 \int_{\xi_h}^{\eta_h} \frac{x^j}{y(x, h)}\, dx, \tag{3.5}$$

where $y(x, h) \geq 0$ is determined isolating y from $H(x, y) = h$. Hence, we have that $\lim_{h \to -2/3+} \frac{I_1(h)}{I_0(h)} = -1$, and we may define the function

$$P(h) = \begin{cases} \dfrac{I_1(h)}{I_0(h)}, & h \in (-2/3, 2/3], \\ -1, & h = -2/3. \end{cases} \tag{3.6}$$

Then (3.4) can be written as

$$I(h) = I_0(h)(\alpha + P(h)).$$

We will prove that P is an increasing function, $P'(h) > 0$ for $h \in (-2/3, 2/3)$, implying that the Abelian integral $I(h)$ will have at most one zero.

Lemma 3.1. *The Abelian integrals $I_0(h)$ and $I_1(h)$ satisfy the Picard–Fuchs equation*

$$(9h^2 - 4)\frac{d}{dh}\begin{pmatrix} I_0 \\ I_1 \end{pmatrix} = \begin{pmatrix} \frac{15h}{2} & 7 \\ \frac{21h}{5} & \frac{21h}{2} \end{pmatrix}\begin{pmatrix} I_0 \\ I_1 \end{pmatrix}. \tag{3.7}$$

Proof. By using (3.5) and the fact that $y^2 = 2h + \frac{2}{3}x^3 - 2x$ along γ_h, we find the recurrence relation

$$I_j(h) = \int_{\gamma_h} \frac{x^j y^2}{y}\, dx = 2hI_j'(h) - 2I_{j+1}'(h) + \frac{2}{3}I_{j+3}'(h). \tag{3.8}$$

On the other hand, using the formula of integration by parts and the fact that $ydy = (-1 + x^2)dx$ along γ_h, we get

$$I_j(h) = \frac{1}{j+1}(I_{j+1}'(h) - I_{j+3}'(h)). \tag{3.9}$$

Removing $I_{j+3}'(h)$ from (3.8) and (3.9), we obtain

$$(2j + 5)I_j(h) = 6hI_j'(h) - 4I_{j+1}'(h).$$

Taking $j = 0, 1$, we have

$$\begin{aligned} 5I_0(h) &= 6hI_0'(h) - 4I_1'(h), \\ 7I_1(h) &= 6hI_1'(h) - 4I_2'(h). \end{aligned} \tag{3.10}$$

Note that along γ_h holds $y^2 dy = (-1 + x^2)ydx$, which implies $I_2(h) \equiv I_0(h)$. The proof follows using $I_0'(h)$ instead of $I_2'(h)$ in (3.10), and solving $I_0'(h)$ and $I_1'(h)$ from this equation. $\qquad\square$

Theorem 3.2. *The function $P(h)$ defined in equation (3.6) is strictly increasing for $h \in (-2/3, 2/3)$.*

Proof. By (3.6) we have

$$P'(h) = \frac{I_1'(h)}{I_0(h)} - \frac{I_0'(h)}{I_0(h)} P(h).$$

Substituting (3.7) into the above equality, we obtain the Riccati differential equation

$$(9h^2 - 4)P' = -7P^2 + 3hP + 5,$$

which is equivalent to the system

$$\begin{aligned}\frac{dh}{dt} &= 9h^2 - 4, \\ \frac{dP}{dt} &= -7P^2 + 3hP + 5.\end{aligned} \tag{3.11}$$

This system has the invariant straight lines $\{h = \pm 2/3\}$, and all four singularities of the system are located on these two lines: a saddle at $S_-(-2/3, -1)$ and a node at $N_-(2/3, -5/7)$ on the lower half plane while a saddle at $S_+(2/3, 1)$ and a node at $N_+(-2/3, 5/7)$ on the upper half plane. Definition (3.6) shows that the graph of the function $P = P(h)$ is the stable manifold of the saddle S_-,

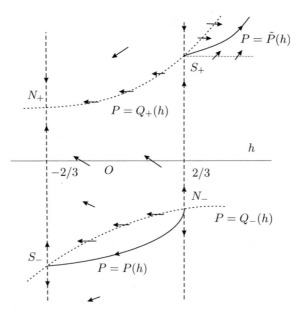

Figure 5: The behavior of the vector field (3.11) and of the function $P(h)$.

and it must go to the unstable node N_- as h increases, since the vector field is upwards on the line $\{(h, P) : P = 0\}$, see Figure 5 (we need only the lower part of this figure in this proof, but we need the upper part in a proof of a lemma in Section 3.3). On the other hand, from the second equation of (3.11) one finds that the horizontal isocline is given by the hyperbola with two branches $P = Q_\pm(h)$, where $Q_\pm(h) = (3h \pm \sqrt{9h^2 + 140})/14$. These two branches divide the strip $\{(h, P) : -2/3 < h < 2/3\}$ into three regions, and the vector field (3.11) is downwards on the top and lower regions, and upwards in the middle region. A direct calculation shows that the slope of the curve $P = P(h)$ at the point S_- is $1/8$ while the slope of $P = Q_-(h)$ at the same point is $1/4$. Hence the curve $P = P(h)$ is located below $P = Q_-(h)$ near this point, therefore it remains below $P = Q_-(h)$ for all $h \in (-2/3, 2/3)$, see again Figure 5. This implies $P'(h) > 0$, since $dP/dt < 0$ below the branch $P = Q_-(h)$ and $dh/dt < 0$ for all $h \in (-2/3, 2/3)$. $\qquad\square$

Remark 3.3. Since equation (3.11) has only regular singularities on $h = \pm 2/3$, it is of *Fuchsian type* (see a detailed proof in Lemma 3.14). A Fuchsian equation is said to be of *Picard–Fuchs type*, provided that it possesses a fundamental set of solutions which are Abelian integrals, see [117] for example.

3.2   A Direct Method

To prove the monotonicity of the function $P = P(h)$, defined as a ratio of two Abelian integrals, one may hope to find some direct ways. For example to construct an auxiliary function defined directly from the Hamiltonian and the integrand functions, some property of the auxiliary function will give the monotonicity of the ratio of the two Abelian integrals. The following result from [198] shows this method under certain restriction on the form of the Hamiltonian.

We first consider the Hamiltonians with the form

$$H(x, y) = \Phi(x) + \Psi(y), \tag{3.12}$$

where $\Phi \in C^2[\alpha, A]$ and $\Psi \in C^2[\beta, B]$. Denote by $\{\gamma_h\}$ the continuous family of ovals contained in the level curves $\{(x, y) : H(x, y) = h, h_1 < h < h_2\}$. Assume that there exist an $a \in (\alpha, A)$ and a $b \in (\beta, B)$, such that the following hypothesis is satisfied:

> (i) $\Phi'(x)(x - a) > 0 \,(\text{or} \ < 0)$ for $x \in (\alpha, A)\backslash\{a\}$,
>
> (ii) $\Psi'(y)(y - b) > 0 \,(\text{or} \ < 0)$ for $y \in (\beta, B)\backslash\{b\}$. $\qquad (H_1)$

This hypothesis implies that the point (a, b) is the center of X_H. Suppose that for each $h \in (h_1, h_2)$, γ_h cuts the line $\{y = b\}$ at the points $(\alpha(h), b)$ and $(A(h), b)$ and cuts the line $\{x = a\}$ at the points $(a, \beta(h))$ and $(a, B(h))$ respectively, where $\alpha \le \alpha(h) \le a \le A(h) \le A$, $\beta \le \beta(h) \le b \le B(h) \le B$. Then for each $x \in (\alpha(h), a)$, there exists a one-to-one mapping $x \mapsto \tilde{x} \in (a, A(h))$, such that $\Phi(x) = \Phi(\tilde{x})$; and for each $y \in (\beta(h), b)$, there exists a one-to-one mapping $y \mapsto \tilde{y} \in (b, B(h))$, such that $\Psi(y) = \Psi(\tilde{y})$.

Now we consider the ratio of the two Abelian integrals

$$F(h) = \frac{I_2(h)}{I_1(h)}, \quad I_k(h) = \oint_{\gamma_h} f_k(x)g(y)dx,$$

where $f_k(x) \in C^1(\alpha, A)$, for $k = 1, 2$, and $g(y) \in C^2(\beta, B)$.

The hypothesis (H_1) implies that $\Phi(x)\Phi'(\tilde{x}) < 0$ and $\Psi(y)\Psi'(\tilde{y}) < 0$ for $x \in (\alpha(h), a)$ and $y \in (\beta(h), b)$. We define two auxiliary functions:

$$\xi(x) = \frac{f_2(x)\Phi'(\tilde{x}) - f_2(\tilde{x})\Phi'(x)}{f_1(x)\Phi'(\tilde{x}) - f_1(\tilde{x})\Phi'(x)}, \quad \eta(y) = \frac{(g(\tilde{y}) - g(y))\Psi'(\tilde{y})\Psi'(y)}{g'(\tilde{y})\Psi'(y) - g'(y)\Psi'(\tilde{y})},$$

where $\tilde{x} = \tilde{x}(x)$ and $\tilde{y} = \tilde{y}(y)$ are defined above.

In order to guarantee that the denominators of $\xi(x)$ and $\eta(y)$ are different from zero, we need two more hypotheses:

$$f_1(x)f_1(\tilde{x}) > 0 \text{ for } x \in (\alpha, a); \tag{H_2}$$

$$g'(y)g'(\tilde{y}) > 0 \text{ for } y \in (\beta, b). \tag{H_3}$$

Remark that the hypothesis (H_2) implies $I_1(h) \neq 0$. If $f_2(x)f_2(\tilde{x}) > 0$ for $x \in (\alpha, a)$, we may use the ratio $I_1(h)/I_2(h)$ instead of $I_2(h)/I_1(h)$.

Theorem 3.4 ([198]). *Assume that $H(x, y)$ has the form (3.12), and the hypotheses (H_1), (H_2), and (H_3) are satisfied, then $\xi'(x)\eta'(y) > 0$ (resp. < 0) for $x \in (\alpha, a)$ and $y \in (\beta, b)$ implies $F'(h) > 0$ (resp. < 0) for $h \in (h_1, h_2)$.*

Proof. We just give a main idea of the proof. Consider

$$I_2'(h)I_1(h) - I_1'(h)I_2(h)$$
$$= \int_{\alpha(h)}^{A(h)} f_2(x) \left(\frac{g'(\tilde{y}(x))}{\Psi'(\tilde{y}(x))} - \frac{g'(y(x))}{\Psi'(y(x))} \right) dx \int_{\alpha(h)}^{A(h)} f_1(x)(g(\tilde{y}(x)) - g(y(x))) \, dx$$
$$- \int_{\alpha(h)}^{A(h)} f_1(x) \left(\frac{g'(\tilde{y}(x))}{\Psi'(\tilde{y}(x))} - \frac{g'(y(x))}{\Psi'(y(x))} \right) dx \int_{\alpha(h)}^{A(h)} f_2(x)(g(\tilde{y}(x)) - g(y(x))) \, dx$$
$$= \frac{1}{2} \int_{\alpha(h)}^{A(h)} \int_{\alpha(h)}^{A(h)} (G(x_1, x_2) + G(x_2, x_1)) \, dx_1 \, dx_2,$$

where $y = y(x)$ and $y = \tilde{y}(x)$ are the two branches of γ_h, and

$$G(x_1, x_2) = (g(\tilde{y}(x_2) - g(y(x_2))) \left(\frac{g'(\tilde{y}(x_1))}{\Psi'(\tilde{y}(x_1))} - \frac{g'(y(x_1))}{\Psi'(y(x_1))} \right)$$
$$\cdot (f_2(x_1)f_1(x_2) - f_2(x_2)f_1(x_1)).$$

By using the definition of $\tilde{x}(x)$ and $\tilde{y}(y)$, and transforming the integration limits of the double integral from $[\alpha(h), A(h)] \times [\alpha(h), A(h)]$ to $[\alpha(h), a] \times [\alpha(h), a]$, we can get the result. □

Now let us use this theorem for the same problem discussed in the last section. Recall that we need to prove the monotonicity of the ratio of the two Abelian integrals $P(h) = I_2(h)/I_1(h)$ defined in (3.4). The Hamiltonian is given in (3.1). Hence in this case we have $\Phi(x) = x - x^3/3$ and $\Psi(y) = y^2/2$; $(a,b) = (-1,0)$; $f_1(x) \equiv 1$ and $f_2(x) = x$; $g(y) = y$. Therefore, all the conditions in (H_1)–(H_3) are satisfied. And it is obvious that $\tilde{y}(y) = -y$, which implies $\eta(y) = y^2$ and $\eta'(y) = 2y < 0$, since $y < 0 < \tilde{y}$. On the other hand, it is easy to compute

$$\xi(x) = \frac{(1 + x\tilde{x})}{x + \tilde{x}}.$$

Note that (H_1) implies $d\tilde{x}/dx = \Phi'(x)/\Phi'(\tilde{x}) < 0$, hence

$$\xi'(x) = \frac{1}{(x + \tilde{x})^2}\left((\tilde{x}^2 - 1) + (x^2 - 1)\frac{d\tilde{x}}{dx} \right) < 0,$$

since $x < -1 < \tilde{x} < 1$. By Theorem 3.4, we obtain $P'(h) > 0$.

Next, we extend the above result to Hamiltonians of the form

$$H(x,y) = \phi(x)y^2 + \Phi(x), \tag{3.13}$$

where $\phi(x), \Phi(x) \in \mathcal{C}^1$ and $\phi(x)$ has a fixed sign. Without loss of generality, we may assume $\phi(x) > 0$. Consider the ratio of the Abelian integrals

$$K(h) = \frac{\displaystyle\int_{\gamma_h} f_2(x)y\,dx}{\displaystyle\int_{\gamma_h} f_1(x)y\,dx},$$

where γ_h is the same as before, and $f_1(x)$ and $f_2(x)$ are continuous functions.

If $\Phi(x)$ satisfies the hypothesis (H_1)(i) and $f_1(x)$ satisfies (H_2), then we can define $\tilde{x} = \tilde{x}(x)$ by $\Phi(x) = \Phi(\tilde{x})$ for $x < a < \tilde{x}$ as before, and define

$$\zeta(x) = \frac{f_2(x)\sqrt{\phi(\tilde{x})}\Phi'(\tilde{x}) - f_2(\tilde{x})\sqrt{\phi(x)}\Phi'(x)}{f_1(x)\sqrt{\phi(\tilde{x})}\Phi'(\tilde{x}) - f_1(\tilde{x})\sqrt{\phi(x)}\Phi'(x)},$$

where $\tilde{x} = \tilde{x}(x)$ for $\alpha < x < a$.

Theorem 3.5 ([198]). *Suppose that $H(x,y)$ has the form (3.13), and the hypotheses (H_1)(i) and (H_2) are satisfied, then the increasing (resp. decreasing) of $\zeta(x)$ for $\alpha < x < a$ implies the decreasing (resp. increasing) of $K(h)$ for $h_1 < h < h_2$.*

Remark 3.6. The monotonicity of the two Abelian integrals can give the uniqueness of limit cycles in a codimension 2 bifurcation problem, as shown above. It is also often used in higher codimension problems, for example see [32, 78, 79, 80, 81, 161]. We will use it again in Chapter 4.

Remark 3.7. New general criteria, applicable in some class of problems, to determine the monotonicity of the ratio of two Abelian integrals or to prove that there are at most two simple zeros for the linear combination of three Abelian integrals have been development in last years. We point out the results of C. Liu, G. Chen & Z. Sun [222] and C. Liu & D. Xiao [224].

In 2011 M. Grau, F. Mañosas & J. Villadelprat [129] generalized the result of [198] from two-dimension of basic Abelian integrals to higher dimension by using the Chebyshev criterion. As in [198] the advantage of this method is that to study the number of zeros of an Abelian integral one only needs to make some purely algebraic computations, unlike the usual way to make complicated differential and integral computations. But this method can be used, up to now, only for restricted forms of the first integrals, like (3.12) or (3.13), where H is an analytic function in some open subset of the plane that has a local minimum at the origin. Then there exists a punctured neighborhood of the origin foliated by ovals $\gamma_h \subset \{H(x,y) = h\}$. If the Abelian integral can be expressed as

$$I(h) = \alpha_0 I_0(h) + \alpha_1 I_1(h) + \cdots + \alpha_{n-1} I_{n-1}(h),$$

where $\alpha_0, \alpha_1, \ldots, \alpha_{n-1}$ are arbitrary constants, then the number of zeros of $I(h)$ is related to the Chebyshev property of the functions $I_0, I_1, \ldots, I_{n-1}$.

Definition 3.8. Let $f_0, f_1, \ldots, f_{n-1}$ be analytic functions on an open interval L of $\mathbb{R}$.

(i) $\{f_0, f_1, \ldots, f_{n-1}\}$ is a *Chebyshev system* (in short, *T-system*) on L if any nontrivial linear combination

$$\alpha_0 f_0(x) + \alpha_1 f_1(x) + \cdots + \alpha_{n-1} f_{n-1}(x)$$

has at most $n - 1$ isolated zeros on L.

(ii) An ordered set of n functions $(f_0, f_1, \ldots, f_{n-1})$ is a *complete Chebyshev system* (in short, *CT-system*) on L if $\{f_0, f_1, \ldots, f_{k-1}\}$ is a T-system for all $k = 1, 2, \ldots, n$.

(iii) An ordered set of n functions $(f_0, f_1, \ldots, f_{n-1})$ is an *extended complete Chebyshev system* (in short, *ECT-system*) on L if, for all $k = 1, 2, \ldots, n$, any nontrivial linear combination

$$\alpha_0 f_0(x) + \alpha_1 f_1(x) + \cdots + \alpha_{k-1} f_{k-1}(x)$$

has at most $k - 1$ isolated zeros on L counted with multiplicities.

We assume that Φ and Ψ are analytic and $x\Phi'(x) > 0$ for any $x \in (x_\ell, x_r)\backslash\{0\}$ and $y\Psi'(y) > 0$ for any $y \in (y_\ell, y_r)\backslash\{0\}$. Then Φ and Ψ must have even multiplicity at 0. Thus, there exist two analytic involutions σ_1 and σ_2 such that

$$\Phi(x) = \Phi(\sigma_1(x)) \text{ for all } x \in (x_\ell, x_r)$$

and

$$\Psi(y) = \Psi(\sigma_2(y)) \text{ for all } y \in (y_\ell, y_r).$$

Note that $\sigma_i(0) = 0$. For a given function κ, we define its balance with respect to σ as

$$\mathscr{B}_\sigma(\kappa)(x) = \kappa(x) - \kappa(\sigma(x)).$$

Theorem 3.9 ([129]). *Consider the Abelian integrals*

$$I_i(h) = \int_{\gamma_h} f_i(x)g(y)\, dx, \text{ for } i = 0, 1, \dots, n-1,$$

where f_i are analytic in $x \in (x_\ell, x_r)$ and g is analytic in $y \in (y_\ell, y_r)$. For each $h \in (0, h_0)$, γ_h is the oval surrounding the origin inside the level curve $\{\Phi(x) + \Psi(y) = h\}$. Let σ_1 and σ_2 be the involutions associated to Φ and Ψ, respectively. Setting $g_0 = g$, define $g_{i+1} = g_i'/\Psi'$. Then $(I_0, I_1, \dots, I_{n-1})$ is an ECT-system on $(0, h_0)$ if the following hypotheses are satisfied:

(i) $(\mathscr{B}_{\sigma_1}(\frac{f_0}{\Phi'}), \mathscr{B}_{\sigma_1}(\frac{f_1}{\Phi'}), \dots, \mathscr{B}_{\sigma_1}(\frac{f_{n-1}}{\Phi'}))$ *is a CT-system on the interval $(0, x_r)$, and*

(ii) $(\mathscr{B}_{\sigma_2}(g_0), \dots, \mathscr{B}_{\sigma_2}(g_{n-1}))$ *is a CT-system on the interval $(0, y_r)$ and* $\mathscr{B}_{\sigma_2}(g_0)(y) = o(y^{2m(n-2)})$.

If the Hamiltonian function H and the function $g(y)$ have the following forms

$$H(x, y) = A(x) + B(x)y^{2m}, \quad g(y) = y^{2s-1},$$

where $s \in \mathbb{N}$, then H has a local minimum at the origin when, by assumption, $B(0) > 0$ and A has a local minimum at $x = 0$. Thus, as before, there exists an involution σ satisfying $A(x) = A(\sigma(x))$ for all $x \in (x_\ell, x_r)$.

Theorem 3.10 ([129]). *Consider the Abelian integrals*

$$I_i(h) = \int_{\gamma_h} f_i(x)y^{2s-1}\, dx, \text{ for } i = 0, 1, \dots, n-1,$$

where, for each $h \in (0, h_0)$, γ_h is the oval surrounding the origin inside the level curve $\{A(x) + B(x)y^{2m} = h\}$. Let σ be the involution associated to A and define

$$\ell_i = \mathscr{B}_\sigma\left(\frac{f_i}{A'B^{\frac{2s-1}{2m}}}\right).$$

Then $(I_0, I_1, \dots, I_{n-1})$ is an ECT-system on $(0, h_0)$ if $s > m(n-2)$ and $(\ell_0, \ell_1, \dots, \ell_{n-1})$ is a CT-system on $(0, x_r)$.

In application problems, if the condition $s > m(n-2)$ is not satisfied, then the following Lemma is useful.

Lemma 3.11 ([129]). *Let γ_h be an oval inside the level curve $\{A(x)+B(x)y^2 = h\}$ and we consider a function F such that F/A' is analytic at $x = 0$. Then, for any $k \in \mathbb{N}$,*

$$\int_{\gamma_h} F(x)y^{k-2}\, dx = \int_{\gamma_h} G(x)y^k\, dx,$$

where $G(x) = \frac{2}{k}\left(\frac{BF}{A'}\right)'(x) - \left(\frac{B'F}{A'}\right)(x)$.

By using Theorem 3.10 the authors of [129] also gave simpler proofs for some know results, including the results in [263] and [385], as well as some new results about cyclicity problems of quadratic integrable systems. Besides, by using Theorem 3.9, the authors of [336] studied perturbations of a class of hyperelliptic Hamiltonian systems with one nilpotent saddle.

In 2011 F. Mañosas & J. Villadelprat [252] generalized the result of Theorem 3.10 to the case that for any nontrivial linear combination of $n - 1$ Abelian integrals to have at most $n - 1 + k$ zeros counted with multiplicities, i.e. when $(I_0, I_1, \ldots, I_{n-1})$ is a *Chebyshev system with accuracy k*. We notice that additional conditions should be added in order that these k extra zeros can appear in the span of the set $(I_0, I_1, \ldots, I_{n-1})$. In [259] is proved that when $k = 1$ always exists an element in the span having this extra zero.

Theorem 3.12 ([252]). *Consider the Abelian integrals*

$$I_i(h) = \int_{\gamma_h} f_i(x)y^{2s-1}\, dx, \ \text{for } i = 0, 1, \ldots, n - 1,$$

where, for each $h \in (0, h_0)$, γ_h is the oval inside the level curve $\{A(x)+B(x)y^{2m} = h\}$. Let σ be the involution associated to A and define

$$\ell_i = \mathscr{B}_\sigma\left(\frac{f_i}{A'B^{\frac{2s-1}{2m}}}\right).$$

If the following conditions are verified (where $W[\ell_0, \ldots, \ell_i]$ stands for the Wronskian determinant of the functions $\ell_0, \ldots, \ell_i$):

(i) *$W[\ell_0, \ldots, \ell_i]$ is non-vanishing on $(0, x_r)$ for $i = 0, 1, \ldots, n - 2$,*

(ii) *$W[\ell_0, \ldots, \ell_{n-1}]$ has k zeros on $(0, x_r)$ counted with multiplicities, and*

(iii) *$s > m(n + k - 2)$,*

then any nontrivial linear combination of $I_0, I_1, \ldots, I_{n-1}$ has at most $n - 1 + k$ zeros on $(0, h_0)$ counted with multiplicities.

As an application of Theorem 3.12, the authors of [252] gave a simple proof about the number of zeros of the Abelian integral corresponding to one period annulus in [82].

We finish this section coming back to the perturbation of reversible quadratic family Q_3^R, i.e. (2.10). We will reinterpret some results in [161] using the above concept of Chebyshev system with accuracy. Moreover, also in [161], the existence of a triple Hopf point is used to study the local cyclicity in the labelled family (r1) in Conjecture 2.23. We will use this bifurcation to improve some of the cyclicity results for the reversible family Q_3^R.

The first Melnikov function (2.13) can be written as the sum of three basic Abelian integrals I_0, I_1, I_2. From the expression of the level curves (2.12), these integrals, interchanging x and y, write similarly as above in Theorem 3.12. As usual, the first Abelian integral I_0 is defined to be the area of the ovals γ_h and, as it is non-vanishing for $h \in (0, h_0)$, we can divide the three basic integrals by this one. These quotients define two new functions, usually denoted by P, Q, that allow us to draw a piece of the curve $(P(h), Q(h))$ in the (P, Q)-plane, because $h \in (0, h_0)$, that we denote by Ω. So, instead of zeros of $M(h)$ we can look for intersection points of Ω with any straight line. This approach is used in many problems of quadratic perturbations to prove that the cyclicity is three because Ω has only one inflexion point. We detail it for the perturbation of quadratic Hamiltonian family in Chapter 4.

The existence of at most one inflexion point is also used in reversible family (2.10) for $\alpha = 3$ in [161] to prove that each one of the two period annuli has cyclicity exactly three, but when both are taken into account only the configuration $(1, 1)$ exists. An interesting comment is that this strategy for bounding the cyclicity of the full period annulus is analogous of the study of zeros of Wronskians, as in Theorem 3.12. Because the existence of a unique inflexion point of the curve (P, Q) is equivalent to prove that (I_0, I_1, I_2) is a Chebyshev system with accuracy 1 in the interval of definition. This property follows directly from [161] because $W_0 := W[I_0] = I_0$, $W_1 := W[I_0, I_1] = W_0^2 P'(h)$ are non-vanishing and $W_2 := W[I_0, I_1, I_2] = W_0^{-1} W_1^2 (Q'(h)/P'(h))'$ vanishes exactly ones. Consequently, using [259], the number of zeros of any linear combination of (I_0, I_1, I_2) is at most three and the upper bound is achieved.

A final observation is that, contrary to what happens with a Chebyshev system, the three zeros cannot be chosen in an arbitrary position in the interval of definition, because they depend on where the zero of W_2 is, which is already fixed according to the parameters of the family. But, as it also commented in [161], this inflexion point is moving varying the parameter β.

3.3   The Method Based on the Argument Principle

In this section we introduce a method to study the number of zeros of Abelian integrals which uses the Argument Principle. G.S. Petrov used this method to study the perturbations of elliptic Hamiltonians of degree 3 and degree 4 in a series of papers [274]–[278]. hence in some literature it is called Petrov's Method. We will use this method to study the same Hamiltonian (3.1), but the Abelian integral $I(h)$ is obtained by polynomial perturbations of arbitrary degree n. Part of the proof below is from [376], we will explain it in Remark 3.19.

The main result in this section is the following theorem.

Theorem 3.13 ([275, 276]). *Any nontrivial $I(h)$, the Abelian integral of the polynomial 1-form of degree at most n over the oval (3.2), has at most $n-1$ zeros for h in the interval $(-2/3, 2/3)$.*

We will prove that $I(h)$ can be extended to the following domain

$$D = \mathbb{C} \backslash \{h \in \mathbb{R}, h \geq 2/3\}$$

as a single-valued analytic function We still use $I(h)$ for the extended complex function in D. In order to apply the Argument Principle to $I(h)$, we define $G = G_{R,\varepsilon} \subset D$ (a simply connected region) being $\partial G = C_{R,\varepsilon}$ a simple closed curve,

$$C_{R,\varepsilon} = \{C_R\} \cup \{C_\varepsilon\} \cup \{L_\pm\},$$

where $C_R = \{h \in \mathbb{C}, |h| = R \gg 1\}$, $C_\varepsilon = \{h \in \mathbb{C}, |h - 2/3| = \varepsilon \ll 1\}$, and $L_\pm$ are the upper and lower banks of the cut $\{2/3 \leq h \leq R\}$, see Figure 6.

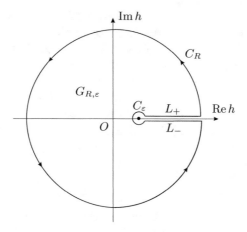

Figure 6: The domain $G_{R,\varepsilon}$ and its boundary.

We first state some technical results, then prove Theorem 3.13 by using them, and finally we give the proofs of the lemmas.

Lemma 3.14. *$I(h)$ can be extended to the domain D as a single-valued analytic function. Moreover, $I_0(h) \sim h^{5/6}$ and $I_1(h) \sim h^{7/6}$ for h in a neighborhood of the infinity.*

Lemma 3.15 ([376]). *Im $I_0(h) \neq 0$ and Im $I_1(h) \neq 0$ for $h \in L_+ \cup L_-$.*

Lemma 3.16 ([376]). *$I_0(h) \neq 0$ for $h \in G \backslash \{-2/3\}$.*

Lemma 3.17. $\operatorname{Im}(I_1(h)/I_0(h)) \neq 0$ for $h \in L_+ \cup L_-$.

Proof of Theorem 3.13. By Lemma 1.13 the Abelian integral $I(h)$ can be expressed in the form

$$I(h) = Q_0(h)I_0(h) + Q_1(h)I_1(h), \tag{3.14}$$

where Q_0 and Q_1 are polynomials with degrees $\deg Q_0 \leq [(n-1)/2] = n_0$ and $\deg Q_1 \leq [n/2] - 1 = n_1$, respectively. The $[\cdot]$ denotes the integer part function. The Abelian integrals $I_j(h)$ are defined in (3.4). Note that $n_0 + n_1 = n - 2$. Here we may suppose that $Q_0(h)$ and $Q_1(h)$ have no common factors.

By Lemma 3.16, $h = -2/3$ is the unique zero of $I_0(h)$, and the limit of the quotient $I_1(h)/I_0(h)$ is -1 as $h \to -2/3^+$ (see (3.6)), hence, instead of $I(h)$ in the above form, we consider the number of zeros of the function

$$F(h) = Q_0 + \frac{I_1}{I_0} Q_1.$$

Note that the trivial zero of $I(h)$ at $h = -2/3$ may be removed for $F(h)$. Now we use the Argument Principle for $F(h)$ to the domain $G_{R,\varepsilon}$, with R and $1/\varepsilon$ positive and big enough. We will prove that the rotation number of F when h turns around the boundary of $G_{R,\varepsilon}$ is at most $n - 1$.

By virtue of Lemma 3.14, when the variable h moves counterclockwise along the curve C_R the function $F(h)$ makes at most $\max(n_0, n_1 + 1/3)$ rotations around zero. By Lemma 3.17 the number of zeros of $\operatorname{Im}(F)$ for $h \in L_+ \cup L_-$ is at most $2n_1$. Since each complete turn of $F(h)$ forces at least two zeros of $\operatorname{Im}(F(h))$ we get that the number of complete turns on these two banks is at most $n_1 + 1$ (we add less than one half turn on each bank). Finally, when h moves along C_ε clockwise, the number of complete turns of F goes to 0, as $\varepsilon \to 0$, since I_1/I_0 tends to a constant as $h \to 2/3^-$, and $Q_0^2(2/3) + Q_1^2(2/3) \neq 0$ ($Q_0(h)$ and $Q_1(h)$ have no common factors). Adding up the above estimate, we get that the total number of rotations is at most $\max(n_0 + n_1 + 1, 2n_1 + 4/3) = n - 1$. Note that the rotation number must be an integer. □

Proof of Lemma 3.14. We know that the Abelian integrals $I_0(h)$ and $I_1(h)$ are real analytic on $h \in (-2/3, 2/3)$. To extend them to a complex domain, we first deduce some differential equations satisfied by them. Rewriting (3.10) as

$$\begin{pmatrix} I_0 \\ I_1 \end{pmatrix} = \begin{pmatrix} \frac{6h}{5} & -\frac{4}{5} \\ -\frac{4}{7} & \frac{6h}{7} \end{pmatrix} \begin{pmatrix} I_0 \\ I_1 \end{pmatrix}', \tag{3.15}$$

and taking derivatives with respect to h on both sides we get

$$\begin{pmatrix} -\frac{1}{5} & 0 \\ 0 & \frac{1}{7} \end{pmatrix} \begin{pmatrix} I_0 \\ I_1 \end{pmatrix}' = \begin{pmatrix} \frac{6h}{5} & -\frac{4}{5} \\ -\frac{4}{7} & \frac{6h}{7} \end{pmatrix} \begin{pmatrix} I_0 \\ I_1 \end{pmatrix}''. \tag{3.16}$$

Substituting (3.16) into (3.15) we obtain

$$I_0 = \frac{4}{5}(4 - 9h^2) I_0'', \quad I_1 = \frac{4}{7}(9h^2 - 4) I_1''. \tag{3.17}$$

It is clear now that if we try to extend h to the complex plane $\mathbb{C}$, then the only singularities of these two equations are located at $h = \pm 2/3$, and the equations are of Fuchsian type because the singularities are regular. It is easy to check that near the point $h = -2/3$ both equations have solutions in power series $\sum_{k=0}^{\infty} a_k (h + 2/3)^k$, which are convergent being $a_k \in \mathbb{C}$. This means that the singularity $h = -2/3$ can be removed, and both Abelian integrals $I_0(h)$ and $I_1(h)$ are holomorphic at this point (we omit the detailed computation). This fact can also be obtained directly from Lemma 20 of [26]. Near the singularity $h = 2/3$, corresponding to the saddle loop Γ of X_H, if we restrict to $h \in (-2/3, 2/3)$ then the solutions of the above equations have the form (see [253, 291]):

$$
\begin{aligned}
I_0(h) &= a_0 + b_0 \, (2/3 - h) \, \ln(2/3 - h) + o((2/3 - h) \ln(2/3 - h)), \\
I_1(h) &= a_1 + b_1 \, (2/3 - h) \, \ln(2/3 - h) + o((2/3 - h) \ln(2/3 - h)),
\end{aligned}
\tag{3.18}
$$

where $a_0 = \oint_\Gamma y dx \neq 0$ and $a_1 = \oint_\Gamma xy dx \neq 0$. A computation shows that in our case $b_0 = b_1 = 1$ and $a_0 b_1 - a_1 b_0 \neq 0$. It is clear that $I_0(h)$ and $I_1(h)$ can not be extended to $\mathbb{C}$ as a single-valued analytic function, unless we cut a ray, starting at the point $h = 2/3$, from $\mathbb{C}$. The first part of the statement is proved. Now we still use the same $I_0(h)$ and $I_1(h)$ for the extended functions on D.

To study the behavior of these two functions near infinity, we let $t = 1/h$ and $K_j(t) = I_j(h)$, then $I_j'' = t^4 \ddot{K}_j + 2t^3 \dot{K}_j$, where $' = d/dh$ and $\dot{} = d/dt$. So, the first equation in (3.17) becomes

$$
t^2 \ddot{K} + 2t \dot{K} + (5/36 + O(t^2)) K = 0.
$$

The index ρ satisfies the equation $\rho(\rho - 1) + 2\rho + 5/36 = 0$, which gives two indices $-1/6$ and $-5/6$. Hence $I_0(h) \sim h^{1/6}$ or $h^{5/6}$ as $h \to \infty$. Similarly, we find from the second equation of (3.17) that $I_1(h) \sim h^{5/6}$ or $h^{7/6}$ as $h \to \infty$. Comparing the coefficients of the leading terms on both sides of (3.15) as $h \to \infty$, we find that the only possibility is $I_0(h) \sim h^{5/6}$ and $I_1(h) \sim h^{7/6}$ as $h \to \infty$. □

Proof of Lemma 3.15. When $h \in L_+ \cup L_-$, equation (3.7) is real analytic, hence it has $(\mathrm{Im}\, I_1, \mathrm{Im}\, I_0)$ as a solution, and the quotient $\mathrm{Im}\, I_1 / \mathrm{Im}\, I_0$ also satisfies equation (3.11). By (3.18) we find $\mathrm{Im}\, I_1 / \mathrm{Im}\, I_0 \to 1$ as $h \in L_+ \cup L_-$ and $h \to 2/3^-$, which implies that the graph of $\mathrm{Im}\, I_1 / \mathrm{Im}\, I_0$ is the unstable manifold of the saddle point $S_+(2/3, 1)$ of system (3.11), see Figure 5. By the same analysis as in the proof of Theorem 3.2 we conclude that when $h > 2/3$ this manifold must stay between the horizontal isocline $P = Q_+(h)$ (the upper branch of the hyperbola) and the ray $\{(h, P) : P = 1, h \geq 2/3\}$, along which the vector field (3.11) is upwards. Hence, we obtain that $1 < \mathrm{Im}\, I_1 / \mathrm{Im}\, I_0 < \infty$ for $h \in L_+ \cup L_-$. □

Proof of Lemma 3.16. We use the same procedure as in the proof of Theorem 3.13 to consider the rotation number of $I_0(h)$ when h turns around the boundary of $G_{R,\varepsilon}$. By using Lemmas 3.14 and 3.15, one can easily obtain that the total rotation number is not bigger than $5/6 + 1$. Since this number must be an integer, $I_0(h)$

has at most one zero in G. As we already have $I_0(-2/3) = 0$, therefore $I_0(h)$ has no other zeros in G.   □

Proof of Lemma 3.17. We suppose the contrary: there exists an $h^* \in L_+ \cup L_-$ such that $\mathrm{Im}(I_1(h^*)/I_0(h^*)) = 0$, which implies

$$\mathrm{Re}(I_0(h^*))\,\mathrm{Im}(I_1(h^*)) - \mathrm{Re}(I_1(h^*))\,\mathrm{Im}(I_0(h^*)) = 0.$$

This means the two vectors $(\mathrm{Im}(I_0(h^*)), \mathrm{Im}(I_1(h^*)))$ and $(\mathrm{Re}(I_0(h^*)), \mathrm{Re}(I_1(h^*)))$ would be proportional. Note that for $h \in L_+ \cup L_-$, these two vector functions are solutions of the real linear differential equation (3.15), hence, being proportional at one point, they are proportional on the entire banks of the cut, i.e. $I_1(h)/I_0(h)$ is real for $h \in L_+ \cup L_-$. From (3.18) we find that for $h \in L_+ \cup L_-$ and h near $2/3$,

$$\mathrm{Im}(I_1(h)/I_0(h)) \sim c(a_0 b_1 - a_1 b_0)(2/3 - h),$$

where c is a non-zero real number. This gives a contradiction.   □

Remark 3.18. By using the result in [291], P. Mardešić [253] generalized the conclusion about the number of zeros of an Abelian integral like the one studied in this section to the conclusion about the number of limit cycles. That is, the maximal number of limit cycles, including the ones bifurcated from the saddle loop, is also $n - 1$, if the polynomial perturbation is of degree at most n.

Remark 3.19. We follow the basic idea of Petrov's proof, but with some changes. For example, there is a claim without proof in Lemma 6 of [276] that the function $\mathrm{Im}\,I_1(h)$ does not have zeros on the open cut, but this fact is not obvious. So we use some proofs from [376].

In [116] L. Gavrilov & I.D. Iliev studied two-dimensional Fuchsian systems in a general setting: under certain conditions the function space of the Abelian integrals obey the Chebyshev property, and there is no need to use the Argument Principle each time. We briefly introduce their result below.

As in (3.14) we consider functions

$$I(h) = p_1(h)I_1(h) + p_2(h)I_2(h), \quad \text{for } h \in (a, b),$$

where $p_1(h)$ and $p_2(h)$ are polynomials, $I_1(h)$ and $I_2(h)$ are complete Abelian integrals over $\gamma_h \subset H^{-1}(h)$, defined as before, and the vector function $\mathbf{I}(h) = (I_1(h), I_2(h))^\mathsf{T}$ satisfies a two-dimensional first-order Fuchsian system

$$\mathbf{I}(h) = \mathbf{A}(h)\,\mathbf{I}'(h), \quad \text{with } ' = d/dh, \tag{3.19}$$

being $\mathbf{A}(h)$ a first-degree polynomial matrix as in (3.15).

The main assumptions on (3.19) are the following:

(H1) The matrix $\mathbf{A}'$ is constant having real distinct eigenvalues.

(H2) The equation $\det \mathbf{A}(h) = 0$ has real distinct roots h_0, h_1, and the identity

$$\text{trace } \mathbf{A}(h) \equiv (\det \mathbf{A}(h))' \tag{3.20}$$

holds.

(H3) The vector function $\mathbf{I}(h)$ is analytic in a neighborhood of h_0.

The conditions that $\mathbf{A}'$ is a constant matrix and $\det \mathbf{A}(h) = 0$ has distinct roots imply that the singularities of the system

$$\mathbf{I}'(h) = \mathbf{A}^{-1}(h)\,\mathbf{I}(h)$$

(including infinity) are regular, i.e. it is of Fuchsian type. The condition (3.20) implies that the characteristic exponents of (3.19) at h_0 and h_1 are $\{0, 1\}$. In the formulation here it is assumed for definiteness that $h_0 < h_1$. A similar result holds if $h_0 > h_1$. Clearly if $h_0 < h_1$ and the function $\mathbf{I}(h)$ is analytic in a neighborhood of $h = h_0$, then it also possesses an analytic continuation in the complex domain $\mathbb{C}\backslash[h_1, \infty)$, as we proved for the special case of Lemma 3.14.

We reformulate the *Chebyshev property* as follows.

Definition 3.20 ([116]). The real vector space of functions V is said to be *Chebyshev* in the complex domain $G \subset \mathbb{C}$ provided that every function $I \in V\backslash\{0\}$ has at most $\dim V - 1$ zeros in G. V is said to be *Chebyshev with accuracy k in G* if any function $I \in V\backslash\{0\}$ has at most $k + \dim V - 1$ zeros in G.

Definition 3.21 ([116]). Let $I(h)$, with $h \in \mathbb{C}$, be a function, locally analytic in a neighborhood of infinity, and $s \in \mathbb{R}$. Write $I(h) \lesssim h^s$, provided that for every sector S centered at infinity there exists a non-zero constant C_s such that $|I(h)| \leq C_s|h|^s$ for all sufficiently big $|h|$, $h \in S$.

For systems (3.19) satisfying (H1) and (H2), the characteristic exponents at infinity are $-\lambda$ and $-\mu$ where $\hat{\lambda} = 1/\lambda$ and $\hat{\mu} = 1/\mu$ are the eigenvalues of the constant matrix $\mathbf{A}'$. According to (H2), we have that $\lambda + \mu = 2$. Denote $\lambda^* = 2$ if λ is an integer number and $\lambda^* = \max(|\lambda - 1|, 1 - |\lambda - 1|)$ otherwise.

Take $s \geq \lambda^*$ and consider the real vector space of functions

$$V_s = \{I(h) = P(h)I_1(h) + Q(h)I_2(h) : P, Q \in \mathbb{R}[h], I(h) \lesssim h^s\},$$

where $\mathbf{I}(h) = (I_1(h), I_2(h))^{\mathsf{T}}$ is a non-trivial solution of (3.19), holomorphic in a neighborhood of $h = h_0$. As $\lambda, \mu \notin \{0, 1, 2\}$ the vector function $\mathbf{I}(h)$ is uniquely determined, up to multiplication by a constant, and $I_1(h_0) = I_2(h_0) = 0$. Clearly, V_s is invariant under linear transformations in (3.19) and affine changes in the argument h. The restriction $s \geq \lambda^*$ is taken to guarantee that V_s is not empty. Recall that $h_0 < h_1$ are the roots of $\det(\mathbf{A}(h)) = 0$.

Theorem 3.22 ([116]). *Assume that conditions (H1)–(H3) hold. If $\lambda \notin \mathbb{Z}$, then V_s is a Chebyshev vector space with accuracy $1 + [\lambda^*]$ in the complex domain $G = \mathbb{C}\backslash[h_1, \infty)$. If $\lambda \in \mathbb{Z}$, then V_s coincides with the space of real polynomials of degree at most $[s]$ which vanish at h_0 and h_1.*

As an application, we use this theorem for the case discussed in the first part of the present section. From (3.15) we have that the matrix in (3.19) is

$$\mathbf{A}(h) = \begin{pmatrix} \frac{6h}{5} & -\frac{4}{5} \\ -\frac{4}{7} & \frac{6h}{7} \end{pmatrix}.$$

It is easy to check that all conditions (H1)–(H3) hold, and $\lambda = 5/6$. Hence, by Theorem 3.22, an upper bound of the number of zeros of $I(h) = Q_0(h)I_0(h) + Q_1(h)I_1(h)$ is $(n-1) + 1 = n$, since the dimension of the function space in this case is n (see Lemma 1.13 and the explanation below it). If we remove the trivial zero at h_0 (in the proof of Theorem 3.13 the zero at $h_0 = -2/3$ is removed), then the upper bound is $n - 1$.

3.4   The Averaging Method

In this section we briefly introduce an application of the averaging method to the study of the weak Hilbert's 16th problem. See more details in [22] and [236]. Although this study is equivalent to the one obtained by using Abelian integrals, in some cases one of them is more convenient than the other. The equivalence of both bifurcation mechanisms is shown in [19, 136]. We first state some general theorems on the averaging method, then use this method to study the quadratic perturbations of a quadratic reversible and non-Hamiltonian system.

The setting of the averaging theory is, in general, in an arbitrary dimensional space. Since we will use it in an autonomous planar system, we will state it only in its one-dimensional form. The following theorem gives a first-order averaging bifurcation mechanism; for a proof we refer the reader to [298].

Theorem 3.23. *Consider the two initial value problems*

$$\dot{x} = \varepsilon f(t, x) + \varepsilon^2 g(t, x, \varepsilon), \quad x(0) = x_0, \tag{3.21}$$

and

$$\dot{y} = \varepsilon f_0(y), \quad y(0) = x_0, \tag{3.22}$$

where $x, y, x_0 \in D$, D is an open subset of $\mathbb{R}$, $t \in [0, \infty)$, $\varepsilon \in (0, \varepsilon_0]$, f and g are periodic of period T in t, and

$$f_0(y) = \frac{1}{T} \int_0^T f(t, y)\, dt. \tag{3.23}$$

Suppose that

(i) *$f, \partial f/\partial x, \partial^2 f/\partial x^2$, and $\partial g/\partial x$ are defined, continuous and bounded by a constant independent of ε in $[0, \infty) \times D$ and $\varepsilon \in (0, \varepsilon_0]$;*

(ii) *T is independent of ε;*

(iii) $y(t)$ belongs to D on the time-scale $1/\varepsilon$.

Then the following statements hold.

(a) On the time-scale $1/\varepsilon$ we have that

$$x(t) - y(t) = O(\varepsilon), \quad as \ \varepsilon \to 0.$$

(b) If p is a singularity of the averaging system (3.22) such that

$$\partial f_0/\partial y|_{y=p} \neq 0, \tag{3.24}$$

then there exists a T-periodic solution $\phi(t, \varepsilon)$ of equation (3.21) which is close to p such that $\phi(t, \varepsilon) \to p$ as $\varepsilon \to 0$.

(c) If (3.24) is negative (resp. positive), then the corresponding periodic solution $\phi(t, \varepsilon)$ in the space (t, x) is asymptotically stable (resp. unstable) for ε sufficiently small.

The next theorem provides a second-order averaging bifurcation mechanism; see [236] or [298] for a proof.

Theorem 3.24. *Consider the two initial value problems*

$$\dot{x} = \varepsilon f(t, x) + \varepsilon^2 g(t, x) + \varepsilon^3 h(t, x, \varepsilon), \quad x(0) = x_0, \tag{3.25}$$

and

$$\dot{y} = \varepsilon f_0(y) + \varepsilon^2 f_{10}(y) + \varepsilon^2 g_0(y), \quad y(0) = x_0, \tag{3.26}$$

with $f, g: [0, \infty) \times D \to G : [0, \infty) \times D \times (0, \varepsilon_0] \to \mathbb{R}$, D an open subset of $\mathbb{R}$, f, g, and h periodic of period T in t, and

$$f_1(t, x) = \frac{\partial f}{\partial x} y_1(t, x) - \frac{\partial y_1}{\partial x} f_0(x),$$

where

$$y_1(t, x) = \int_0^t (f(s, x) - f_0(x)) \, ds + z(x),$$

with $z(x)$ a C^1 function such that the averaging of y_1 is zero. Besides, f_0, f_{10}, and g_0 denote the averaging functions of f, f_1, and g, respectively, defined as in (3.23). Suppose that

(i) $\partial f/\partial x$ is Lipschitz in x and all these functions are continuous on their domain of definition;

(ii) $|h(t, x, \varepsilon)|$ is bounded by a constant uniformly in $[0, L/\varepsilon) \times D \times (0, \varepsilon_0]$;

(iii) T is independent of ε;

(iv) $y(t)$ belongs to D on the time-scale $1/\varepsilon$.

Then

(a) *On the time-scale $1/\varepsilon$ we have that*

$$x(t) = y(t) + \varepsilon y_1(t, y(t)) + O(\varepsilon^2), \quad as \ \varepsilon \to 0.$$

If, in addition, $f_0(y) \equiv 0$, then the following statements hold.

(b) *If p is a singularity of the averaging system (3.26) such that*

$$\frac{\partial}{\partial y}(f_{10}(y) + g_0(y))\Big|_{y=p} \neq 0, \tag{3.27}$$

then there exists a T-periodic solution $\phi(t, \varepsilon)$ of equation (3.25) which is close to p such that $\phi(t, \varepsilon) \to p$ as $\varepsilon \to 0$.

(c) *If (3.27) is negative (resp. positive), then the corresponding periodic solution $\phi(t, \varepsilon)$ in the space (t, x) is asymptotically stable (resp. unstable) for ε sufficiently small.*

Now we consider a perturbation of a planar integrable system of the form

$$X_\varepsilon : \begin{cases} \dot{x} = P(x, y) + \varepsilon\, p(x, y), \\ \dot{y} = Q(x, y) + \varepsilon\, q(x, y), \end{cases} \tag{3.28}$$

where $P, Q, p, q \in \mathcal{C}^1(\mathbb{R}^2, \mathbb{R})$. We suppose that the unperturbed system X_0 has an integrating factor $\mu(x, y) (\neq 0)$, a first integral H and a continuous family of ovals $\{\gamma_h\}$:

$$\gamma_h \subset \{(x, y) : H(x, y) = h, \ h_1 < h < h_2\}.$$

To study the number of limit cycles of system (3.28) for sufficiently small ε by using the above theorems, a natural question is how do we transform this system to the form (3.21) or (3.25). The following result gives an answer.

Theorem 3.25 ([22]). *Assume that $xQ(x, y) - yP(x, y) \neq 0$ for all (x, y) in the period annulus formed by the ovals $\{\gamma_h\}$. Then there is a continuous function $\rho : (\sqrt{h_1}, \sqrt{h_2}) \times [0, 2\pi) \to [0, \infty)$ such that*

$$H(\rho(R, \varphi) \cos \varphi, \ \rho(R, \varphi) \sin \varphi) = R^2,$$

for all $R \in (\sqrt{h_1}, \sqrt{h_2})$ and all $\varphi \in [0, 2\pi)$, and the differential equation which describes the dependence between the square root of energy, $R = \sqrt{h}$, and the angle φ for system (3.28) is

$$\frac{dR}{d\varphi} = \varepsilon \frac{\mu\,(x^2 + y^2)(Qp - Pq)}{2R(Qx - Py)}\left(1 - \varepsilon\,\frac{qx - py}{Qx - Py}\right) + O(\varepsilon^3),$$

where $x = \rho(R, \varphi) \cos \varphi$ and $y = \rho(R, \varphi) \sin \varphi$.

Example 3.26 ([22]). Consider

$$\dot{x} = -y + x^2 + \varepsilon\, p(x, y),$$
$$\dot{y} = x + xy + \varepsilon\, q(x, y), \tag{3.29}$$

where $p(x, y) = a_1 x - a_3 x^2 + (2a_2 + a_5)xy + a_6 y^2$ and $q(x, y) = a_1 y + a_2 x^2 + (2a_3 + a_4)xy - a_2 y^2$. We will show that, for $\varepsilon \approx 0$, system (3.29) can have two limit cycles bifurcating from any two different level curves of the full period annulus.

Note that when $\varepsilon = 0$ system (3.29) is reversible and non-Hamiltonian (the center is isochronous), which has the first integral

$$H(x, y) = \frac{x^2 + y^2}{(1 + y)^2}$$

and the corresponding integrating factor is $2(1 + y)^{-3}$. We use Theorem 3.25 and taking the transformation $x = \rho \cos\varphi, y = \rho \sin\varphi$, where

$$\rho = \rho(R, \varphi) = \frac{R}{1 - R\sin\varphi}, \quad \text{for } 0 < R < 1, \ \varphi \in [0, 2\pi),$$

then system (3.29) becomes

$$\frac{dR}{d\varphi} = \varepsilon\, \frac{a_1 R + a(\varphi)R^2 + b(\varphi)R^3}{1 - R\sin\varphi} + O(\varepsilon^2), \tag{3.30}$$

where

$$a(\varphi) = -(2a_1 + a_2)\sin\varphi + (2a_3 + a_4 + a_6)\cos\varphi$$
$$(4a_2 + a_5)\sin\varphi\cos^2\varphi - (3a_3 + a_4 + a_6)\cos^3\varphi, \tag{3.31}$$
$$b(\varphi) = a_1 + a_2 - (2a_3 + a_4)\sin\varphi\cos\varphi - (a_1 + 2a_2)\cos^2\varphi.$$

By integration one has the averaging function

$$f_0(R) = \frac{1}{2\pi} \int_0^{2\pi} \frac{a_1 R + a(\varphi)R^2 + b(\varphi)R^3}{1 - R\sin\varphi}\, d\varphi$$
$$= \frac{-1}{2R\sqrt{1 - R^2}} [2a_2 R^4 - (2a_1 - 6a_2 - a_5)R^2\sqrt{1 - R^2} \tag{3.32}$$
$$- (10a_2 + 2a_5)R^2 - (8a_2 + 2a_5)\sqrt{1 - R^2} + 8a_2 + 2a_5].$$

Note that $R \in (0, 1)$. By the substitution $\xi = \sqrt{1 - R^2}$ the numerator of the above expression write as

$$\xi(1 - \xi)(2a_2\xi^2 + (2a_1 - 4a_2 - a_5)\xi + 2a_1 + 2a_2 + a_5).$$

As the last factor writes as a general quadratic polynomial $b_2\xi^2 + b_1\xi + b_0$, with the linear change $(a_1, a_2, a_5) = ((b_0 + b_1 + b_2)/4, b_2/2, (b_0 - b_1 - 3b_2)/2)$, the first averaging function $f_0(R)$ has at most two zeros for $R \in (0,1)$. Moreover, there exist values of the perturbation parameters such that it has exactly two simple zeros. Hence, up to a first-order analysis, the system (3.29) has at most two limit cycles for ε sufficiently small and the upper bound is reached for an specific perturbation.

□

Remark 3.27. The computations of the above example are quite different from the one in the first edition of this text. We have recovered the original change in [22] and we have corrected a misprint in the polynomial perturbation. The result in [22] was obtained for quadratic perturbations in its Bautin normal form but with the same misprint in the writing of the perturbation function q.

The key point of the above computations is that the integral (3.32) can be obtained explicitly. This is due to the fact that the denominator is a trigonometrical polynomial of degree one. This advantage, together with the existence of a birational linearization, was used in [284] to study the second-order averaging function in the above and some other quadratic isochronous systems showing that no more than two limit cycles appear. In fact, the unperturbed ($\varepsilon = 0$) system (3.29) is the quadratic isochronous center S_2, as we have already commented previously, that linearizes with the transformation $(u,v) = (x/(1+y), y/(1+y))$. With this change, (3.29) writes as the next specific rational perturbation of the linear center

$$
\dot{u} = -v + \frac{\varepsilon}{1-v}(a_1 u - a_3 u^2 - (2a_1 - 2a_2 - a_5)uv + a_6 v^2
$$
$$
- a_2 u^3 + (a_1 + a_2)uv^2 - (2a_3 + a_4)u^2 v),
$$
$$
\dot{v} = u + \varepsilon(a_1 v + a_2 u^2 + (2a_3 + a_4)uv - (a_1 + a_2)v^2).
$$

With the usual change to polar coordinates $(u,v) = (R\cos\varphi, R\sin\varphi)$, the corresponding Abelian integral writes as $\hat{f}_0(R) = -Rf_0(R)$, which provides the same conclusion as in Example 3.26.

These explicit expressions of the integrals can be used also for obtaining the "shape" of the limit cycles, writing the first terms of the Taylor series in ε of them in the form $R(\varphi,\varepsilon) = R_0 + R_1(\varphi)\varepsilon + \cdots$, see more details in [282, 283]. The main difficulty is that the first-order study in ε only provides the value of R_0, while to know the expression of $R_1(\varphi)$ a second-order study is necessary. More concretely, only with a first-order analysis the value of $R_1(0) = \rho_1$ can not be obtained. We will illustrate this property in the next example, computing the first terms of the Taylor development with respect to ε of a solution of (3.29). We will closely follow the scheme developed in [283]. For a relation between the first-order of the Taylor development of the Lyapunov constants with the first order of averaging see [128].

Example 3.28. Let $(a_1, a_2, a_3, a_4, a_5, a_6) = (9/50, 1, 0, 0, -91/25, 0)$ be the perturbation parameters in system (3.29). Then, (3.32) writes as

$$
f_0(R) = \frac{-25R^4 - 25R^2\sqrt{1-R^2} + 34R^2 + 9\sqrt{1-R^2} - 9}{25R\sqrt{1-R^2}}
$$

with $f_0(3/5) = 0$ and $f_0'(3/5) = -2/5$. Hence, there exists a periodic solution of (3.30) that writes as

$$R(\varphi, \varepsilon) = 3/5 + R_1(\varphi)\varepsilon + R_2(\varphi)\varepsilon^2 + \cdots . \tag{3.33}$$

The value of $\rho_0 = 3/5$ could be also computed directly looking for a periodic solution of (3.30) of the form $R(\varphi, \varepsilon) = \rho_0 + R_1(\varphi)\varepsilon + \cdots$, equating the terms in ε and imposing the periodicity in $R_1(\varphi)$. This first-order analysis allow us to write

$$R_1(\varphi) = \rho_1 - \frac{531 - (531 - 135\sin\varphi)\cos\varphi}{1250},$$

but the periodicity of R_1 does not provides any information about the initial condition ρ_1. This fact explains why, as we have commented above, the complete knowledge of the first-order approximation needs the second-order analysis. Instead of introducing here complicated expressions of higher order averaging analysis, we have used the direct computation of the functions R_1 and R_2 substituting (3.33) in (3.30). The function R_1 follows equating the terms in ε and integrating in φ. The value of $\rho_1 = 531/1250$ is obtained looking at the terms in ε^2 and imposing that $R_2(\varphi)$ is 2π-periodic. Similarly, we can obtain

$$R_2(\varphi) = \rho_2 - \frac{729}{25000}\sin^4\varphi + \frac{4779}{31250}\sin^3\varphi - \frac{181899}{625000}\sin^2\varphi$$
$$- \frac{39429}{62500}\sin\varphi - \frac{1632}{3125}\ln\left(\frac{5 - 3\sin\varphi}{5}\right).$$

Finally, from the third-order analysis, we get

$$\rho_2 = -\frac{1632}{3125}(25\ln 2 - 18\ln 3 + \ln 5) - \frac{65079}{625000}.$$

In Figure 7 left we have drawn the initial conditions $R(0) = \rho_0$ obtained numerically for $\varepsilon \in (-0.1, 0.2)$ together with the second-order approximation in ε obtained above. For $\varepsilon = 0.2$, the numerical periodic solution versus the second-order approximation are depicted in Figure 7 right, respectively with solid and dot curves. □

We notice that in the last example, we have chosen an specific perturbation instead of a generic one to get simpler computations. But the difficulties appear when we increase the order of approximation in ε.

3.5 The Averaging Method in Piecewise Systems

In last years, the averaging method has been used also to study the existence of periodic orbits in equations of type (3.25) but defined by piecewise functions as, for example,

$$\dot{x} = \begin{cases} \varepsilon f^+(t, x) + \varepsilon^2 g^+(t, x, \varepsilon), & \text{for } t \in [0, T/2], \\ \varepsilon f^-(t, x) + \varepsilon^2 g^-(t, x, \varepsilon), & \text{for } t \in [T/2, T]. \end{cases} \tag{3.34}$$

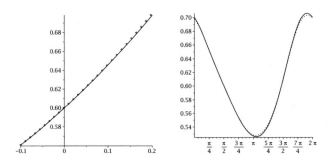

Figure 7: The initial conditions $R(0)$ for $\varepsilon \in (-0.1, 0.2)$ and the periodic solution for $\varepsilon = 0.2$ in Example 3.28.

In [238], Theorem 3.23 is extended to differential equations of type (3.34) being the first averaging function the corresponding piecewise version of f_0 in (3.23), that is

$$f_0(y) = \frac{1}{T}\left(\int_0^{T/2} f^+(t,y)dt + \int_{T/2}^T f^-(t,y)dt \right). \tag{3.35}$$

Hence, all the problems and techniques explained in this work can be considered in this research line. But as this is not the aim of this work, we only observe a new extension of the notion of Abelian integral. We show the similarities and differences with the classical approach just with an example: the piecewise version of the perturbed system (3.29). That is, a piecewise quadratic perturbation of a particular quadratic isocronous center. For a complete description of the first and second-orders studies for the quadratic isocronous centers having a birational linearization we refer the reader to [69].

Example 3.29. Consider

$$\begin{aligned} \dot{x} &= -y + x^2 + \varepsilon\, p^{\pm}(x,y), \\ \dot{y} &= x + xy + \varepsilon\, q^{\pm}(x,y), \end{aligned} \tag{3.36}$$

where $p^{\pm}(x,y) = a_1^{\pm} x - a_3^{\pm} x^2 + (2a_2^{\pm} + a_5^{\pm})xy + a_6^{\pm} y^2$ and $q^{\pm}(x,y) = a_1^{\pm} y + a_2^{\pm} x^2 + (2a_3^{\pm} + a_4^{\pm})xy - a_2^{\pm} y^2$. Here the perturbation (p^+, q^+) is defined in $y > 0$ and (p^-, q^-) in $y < 0$. We will show that, for $\varepsilon \approx 0$, system (3.36) can have 5 limit cycles bifurcating from any 5 different level curves of the full period annulus.

With the transformation used in Example 3.26, we have that the piecewise system (3.36) becomes

$$\frac{dR}{d\varphi} = \varepsilon \frac{a_1^{\pm} R + a^{\pm}(\varphi)R^2 + b^{\pm}(\varphi)R^3}{1 - R\sin\varphi} + O(\varepsilon^2),$$

being $a^{\pm}, b^{\pm}$ the corresponding piecewise functions of the ones introduced in (3.31), respectively defined in $y > 0$ and $y < 0$. Therefore, the averaging function (3.35), after and adequate linear change of coordinates in the parameter space, is

$$f_0(R) = \frac{1}{2\pi}\left(\int_0^\pi \frac{a_1^+ R + a^+(\varphi)R^2 + b^+(\varphi)R^3}{1 - R\sin\varphi}\,d\varphi\right.$$
$$\left. + \int_\pi^{2\pi} \frac{a_1^- R + a^-(\varphi)R^2 + b^-(\varphi)R^3}{1 - R\sin\varphi}\,d\varphi\right)$$
$$= \frac{1}{4\pi R\sqrt{1 - R^2}}\left[((b_4 - b_5)R^4 + b_5 R^2 - b_4)\arctan\left(\frac{R}{\sqrt{1-R^2}}\right)\right.$$
$$\left. + (b_2 R^3 + b_1 R^2 + b_4 R - b_3)\sqrt{1 - R^2} - (b_3 + b_0)R^4 + b_0 R^2 + b_3\right]$$
$$= \frac{1}{4\pi R\sqrt{1 - R^2}}\sum_{k=0}^5 b_k\, h_k(R).$$

We notice that the main difference with the averaging function (3.32) is that now there are six independent functions and the problem of studying its simple zeros, due to the arctan function, is no longer algebraic. The proof finishes proving, using Proposition 1.15, that the ordered set of functions $(h_0, h_1, \ldots, h_5)$ defines an ECT-system on the interval $(0, 1)$.

Straightforward computations give us the expressions of the corresponding Wronskians $W_i = W[h_0, \ldots, h_i]$, for $i = 0, \ldots, 5$, where

$$W_0 = R^2(1 - R^2),$$
$$W_1 = R^5\sqrt{1 - R^2},$$
$$W_2 = -R^6,$$
$$W_3 = \frac{6R^3}{1 - R^2}((R^2 - 4)\sqrt{1 - R^2} - 3R^2 + 4),$$
$$W_4 = \frac{-12R^4}{(1 - R^2)^{5/2}}\left[9R\arctan\left(\frac{R}{\sqrt{1-R^2}}\right)\right.$$
$$\left. + (3R^4 - 22R^2 + 28)\sqrt{1 - R^2} - 12R^4 + 27R^2 - 28\right],$$
$$W_5 = \frac{-72R^5}{(1 - R^2)^{9/2}}\left[3(12R^2 + 5)\sqrt{1 - R^2}\arctan\left(\frac{R}{\sqrt{1-R^2}}\right) + \right.$$
$$\left. - 16R(R^2 + 4)\sqrt{1 - R^2} - R(R^2 - 1)(2R^2 + 49)\right].$$

Clearly, W_0, W_1, W_2 are non-vanishing in $(0, 1)$ and also W_3, after doing the change $R = \sqrt{1 - \xi^2}$ if necessary. The proof for the last two require more work. In order to simplify the computations, we write them, multiplying by an adequate non-zero function, in the form $\widehat{W}_i = \arctan\xi + F_i(\xi)$, for $i = 4, 5$, where $R = \xi(1 + \xi^2)^{-1/2}$ and they are well defined when $\xi > 0$. Now, the first derivative of $\widehat{W}_i$ is a rational function of $\sqrt{1 + \xi^2}$ and ξ. It is easy to check that $\widehat{W}_4'$ has no zeros and $\lim_{\xi \to 0^-} \widehat{W}_4 = 0$, so W_4 is a non-vanishing function. Finally, W_5 has no zeros because $\widehat{W}_5'$ vanishes only ones at $\sqrt{26 + 6\sqrt{21}}/2$, $\widehat{W}_5 = \frac{2}{525}\xi^7 + \cdots$ when $\xi \approx 0$, and $\lim_{\xi \to +\infty} \widehat{W}_5 = \frac{\pi}{2} - \frac{80}{51} > 0$.

$\square$

3.6   Other Methods and Related Works

Before closing this chapter, we remark that the methods introduced here mainly remain in the real domain (except for the use of the Argument Principle). There is also a method based on complexification of the Abelian integrals. The basic construction, see [7] for example, is the following. Consider the Hamiltonian function $H \in \mathbb{C}[x, y]$ as a complex map $\mathbb{C}^2 \to \mathbb{C}$, and define the Abelian integral as an integral of the complex polynomial 1-form over some cycle on the level set $H^{-1}(h)$ (a one-dimensional complex manifold). This permits us to construct the monodromy map, and it is possible to apply the Picard–Lefschetz formula and to use the tools of algebraic topology. We list some works below in which this complex method was successfully applied. In [165] Y. Ilyashenko studied the codimension two Bogdanov–Takens bifurcation (the same problem introduced in this chapter). In [254] P. Mardešić gave an explicit bound for the multiplicity of zeros of Abelian integrals under certain generic assumptions. In [113] L. Gavrilov studied the quadratic perturbation of quadratic Hamiltonian systems (the weak Hilbert's 16th problem for $n = 2$). In [261] D. Novikov & S. Yakovenko gave an explicit upper bound of the polynomial perturbations of the hyperelliptic Hamiltonian under some generic condition. For a more detailed introduction of this complex theory we refer to an early work of Y. Ilyashenko [164], the lecture notes [345], and the book by Y. Ilyashenko & S. Yakovenko [171].

There are numerous works on the weak Hilbert's 16th problem. At the end of the last three chapters, we list some of them, mainly by N.N. Bautin, R.I. Bogdanov, M. Caubergh, H. Chen, X. Chen, C. Chicone, C. Christopher, B. Coll, P. De Maesschalck, F. Dumortier, J.-P. Françoise, A. Gasull, L. Gavrilov, J. Giné, M. Han, X. Hong, E. Horozov, W. Huang, I.D. Iliev, Y. Ilyashenko, B. Li, C. Li, F. Li, J. Li, W. Li, C. Liu, Y. Liu, J. Llibre, P. Mardešić, L. Peng, G.S. Petrov, R. Prohens, J. Reyn, V. Romanovski, R. Roussarie, C. Rousseau, S. Ruan, D. Schlomiuk, J. Sotomayor, F. Takens, Y. Tang, Y. Tian, J. Torregrosa, J. Villadelprat, Y. Xia, D. Xiao, S. Yakovenko, Y. Ye, J. Yu, P. Yu, W. Zhang, X. Zhang, Z. Zhang, L. Zhao, Y. Zhao, H. Zhu, H. Żołądek, and their collaborators:

- About some general studies on Abelian integrals, see [84, 100, 112, 131, 136, 137, 260, 274, 275, 281, 338];

- About quadratic perturbations of quadratic Hamiltonian systems, see [33, 40, 54, 113, 154, 156, 258, 375, 385, 389];

- About higher-order perturbations of quadratic Hamiltonian systems, see [165, 324];

- About quadratic perturbations of quadratic integrable systems, see [65, 82, 159, 161, 188, 195, 220, 221, 263, 264, 267, 268, 269, 270, 271, 308, 341, 359, 393];

- About higher-order perturbations of quadratic integrable systems, see [148, 150];

- About perturbations of cubic Hamiltonian systems, see [118, 142, 152, 162, 187, 205, 207, 209, 215, 234, 313, 314, 315, 325, 340, 348, 351, 357, 362, 386, 387, 394];

- About perturbations of Hamiltonian systems, see [23, 45, 102, 117, 138, 143, 199, 219, 253, 286, 336, 350, 352, 356, 360, 377, 378, 379, 380, 384];

- Related to perturbations in the class of Liénard equations, see [30, 34, 35, 36, 78, 79, 80, 81, 92, 184, 240, 248, 343, 363];

- Related to B–T bifurcations, see [15, 37, 38, 87, 96, 97, 135, 160, 182, 196, 274, 275, 276, 317, 320];

- Related to bifurcations of Z_q symmetry, see [151, 200, 201, 204, 208, 210, 211, 212, 213, 230, 231, 234];

- Related to bifurcations of degenerate singularities, see [21, 123, 133, 202, 322, 361];

- Related to Hopf bifurcations, see [8, 11, 26, 58, 63, 89, 106, 107, 121, 126, 127, 128, 191, 216, 225, 226, 227, 229, 232, 233, 244, 262, 289, 310, 370];

- Related to homoclinic or heteroclinic bifurcations, see [2, 97, 139, 140, 144, 146, 153, 291, 319, 321, 323, 337, 346, 347, 349, 364, 368];

- Related to higher-order Melnikov functions, see [90, 114, 157, 158, 163, 176, 328, 374];

- Related to algebraic limit cycles, see [25, 243, 369];

- Related to Alien limit cycles, see [27, 28, 61, 250];

- Related to monotonicity of two Abelian integrals or Chebyshev property of a class of Abelian integrals, see [101, 129, 116, 179, 222, 223, 252, 256, 257, 276, 339];

- Related to averaging method, see [22, 122, 124, 125, 236, 239, 265, 266, 298, 307];

- Related to period functions, see [17, 42, 43, 44, 46, 48, 49, 50, 55, 60, 68, 93, 95, 99, 104, 105, 111, 192, 255, 290, 329, 330, 331, 332, 335, 381, 382, 383];

- Related to finiteness problem or Hilbert's number, see [57, 74, 75, 76, 77, 85, 86, 88, 134, 174, 285, 295, 296, 327, 391];

- About survey papers related to weak Hilbert's 16th problem, see [31, 167, 180, 183, 203, 303, 311, 312, 345];

- About books related to weak Hilbert's 16th problem, see [4, 53, 130, 132, 141, 145, 169, 171, 214, 235, 237, 251, 293, 354, 355, 373, 376].

Chapter 4

A Unified Proof of the Weak Hilbert's 16th Problem for n=2

4.1 Preliminaries and the Centroid Curve

Any cubic generic Hamiltonian with at least one period annulus contained in its level curves, as we have explained in Subsection 1.2.1, can be transformed into the normal form

$$H(x,y) = \frac{1}{2}(x^2 + y^2) - \frac{1}{3}x^3 + axy^2 + \frac{1}{3}by^3, \tag{4.1}$$

where a, b are parameters lying in the open region

$$G = \left\{ (a,b) : -\frac{1}{2} < a < 1, \ 0 < b < (1-a)\sqrt{1+2a} \right\}. \tag{4.2}$$

We recall that the region G is also defined from the interior of (1.4) and it can be written as the union of three open regions G_1, G_2, and G_3. Figure 1 shows all five possible phase portraits of X_H in the generic cases. That is for values (a,b) in G_1, G_2, G_3, l_2, and l_∞. Here X_H is the Hamiltonian vector field corresponding to H, i.e.

$$X_H = H_y \frac{\partial}{\partial x} - H_x \frac{\partial}{\partial y}.$$

As previously, the subscripts x and y denote the respective first derivatives. The vector field X_H has a center at the origin in the (x,y)-plane, and the continuous family of ovals, surrounding the center, is

$$\{\gamma_h\} \subset \{(x,y) : H(x,y) = h, \ 0 < h < 1/6\}.$$

The oval γ_h shrinks to the center as $h \to 0^+$, and the oval γ_h terminates at the saddle loop of the saddle point $(1,0)$ when $h \to 1/6^-$.

We consider any quadratic perturbation of X_H, i.e.

$$X_\varepsilon = X_H + \varepsilon Y_\varepsilon,$$

where

$$Y_\varepsilon = f(x, y, \varepsilon)\frac{\partial}{\partial x} + g(x, y, \varepsilon)\frac{\partial}{\partial y},$$

with f and g polynomials in x and y of degree 2, and their coefficients depend analytically on the parameter ε.

In this chapter we will study the Abelian integral

$$I(h) = \oint_{\gamma_h} f(x, y, 0)dy - g(x, y, 0)dx, \tag{4.3}$$

and prove the following theorem (see Subsection 1.2.1 and [32]).

Theorem 4.1. *For any cubic polynomial H with $(a, b) \in G$, defined in (4.2), and any quadratic polynomials f and g, the least upper bound for the number of zeros of the Abelian integral (4.3) is 2.*

Since the orientation of the integral over the oval γ_h is clockwise, and f and g are polynomials in x and y of degree 2, we may rewrite, by using Green's formula, the integral (4.3) into the form

$$I(h) = -\iint_{\text{Int } \gamma_h} \left(\frac{\partial f}{\partial x} + \frac{\partial g}{\partial y}\right)\bigg|_{\varepsilon=0} dxdy = \iint_{\text{Int } \gamma_h} (\alpha + \beta x + \gamma y)\, dx\, dy, \tag{4.4}$$

where α, β, and γ are arbitrary constants and Int γ_h denotes the interior of the oval γ_h. Following the notation of [154], we define the functions

$$M(h) = \iint_{\text{Int } \gamma_h} dx\, dy, \quad X(h) = \iint_{\text{Int } \gamma_h} x\, dx\, dy,$$

$$Y(h) = \iint_{\text{Int } \gamma_h} y\, dx\, dy, \quad K(h) = \iint_{\text{Int } \gamma_h} xy\, dx\, dy. \tag{4.5}$$

In the next section we will see why we need to introduce four functions instead of the three that appear in (4.4). Since $M(h)$ is the area of Int γ_h and $M'(h)$ is the period of γ_h, we have that $M(h) > 0$ and $M'(h) > 0$ for $h \in (0, 1/6)$. Hence (4.4) can be written as

$$I(h) = \alpha M(h) + \beta X(h) + \gamma Y(h) = M(h)\left[\alpha + \beta p(h) + \gamma q(h)\right], \tag{4.6}$$

where

$$p(h) = \frac{X(h)}{M(h)}, \quad q(h) = \frac{Y(h)}{M(h)}. \tag{4.7}$$

The following result is easily obtained from the definitions of $I(h)$, $p(h)$, and $q(h)$. The second claim of next statement (iii) was firstly proved in Theorem 2.4 of [154], see also Lemma 4.1 of [199].

Lemma 4.2. *For any $(a, b) \in G$ the functions I, p, and q satisfy the next properties:*

(i) $I(0) = 0$ *for any constants α, β, and γ.*

(ii) $p(0) = \lim\limits_{h \to 0^+} p(h) = 0$ *and* $q(0) = \lim\limits_{h \to 0^+} q(h) = 0$.

(iii) $0 < p(h) < 1$ *and* $q(h) < 0$ *for* $h \in (0, 1/6]$.

(iv) $p, q \in C^\infty [0, 1/6) \bigcup C^0 [0, 1/6]$.

Note that in the (x, y)-plane the point $(p(h), q(h))$ is the coordinate of the center of mass of Int γ_h with uniform density, so following [154] we give the definition below.

Definition 4.3. In the (p, q)-plane, for $(a, b) \in G$, the curve

$$\Sigma_{a,b} = \{(p, q) : p = p(h), q = q(h), \ 0 \le h \le 1/6\}$$

is called a *centroid curve*.

The geometric meaning of this curve is that for any fixed $(a, b) \in G$, the level h gives a point on $\Sigma_{a,b}$, which is the center of mass of Int γ_h if we identify the (p, q)-plane with the (x, y)-plane. Hence, it is natural that $(p(h), q(h)) \to (0, 0)$ as $h \to 0^+$ and $(p(1/6), q(1/6))$ is the center of mass of the region surrounded by the saddle loop $\gamma_{1/6}$.

It is obvious from (4.6) that for any constants α, β, and γ the number of zeros of $I(h)$ for $h > 0$ (counting the multiplicities) equals the number of intersection points (counting the multiplicities) of the curve $\Sigma_{a,b}$ with the straight line

$$L_{\alpha\beta\gamma} = \{(p, q) : \alpha + \beta p + \gamma q = 0\},$$

where $\beta^2 + \gamma^2 \neq 0$.

Definition 4.4. A plane curve is called *sectorial*, if it is smooth, and when running it, the tangential vector rotates through an angle less than π.

If X_H has only one period annulus, then Theorem 4.1 follows from the following result.

Theorem 4.5. *For any $(a, b) \in G$ the curve $\Sigma_{a,b}$ is sectorial, and is strictly convex with non-zero curvature.*

For $(a, b) \in G_2$, the vector field X_H has two period annuli (see Figure 1), hence there are two centroid curves $\Sigma_{a,b}$ and $\hat{\Sigma}_{a,b}$. To finish the proof of Theorem 4.1, we also need the following result, which was first proved in [152] based on some other results, and we will give a direct proof in the last section of this chapter.

Theorem 4.6. *For any $(a, b) \in G_2$ both centroid curves are strictly convex with non-zero curvature, and any straight line cuts $\Sigma_{a,b} \cup \hat{\Sigma}_{a,b}$ at most at two points, counting the multiplicities.*

4.2 Basic Lemmas and the Geometric Proof of the Result

Straightforward computations show that it is impossible to deduce a third-order Picard–Fuchs equation satisfied by $M(h)$, $X(h)$, and $Y(h)$. In fact, it is necessary to add one more function, for example $K(h) = \iint_{\text{Int } \gamma_h} xy\, dx\, dy$. Then one may deduce a Picard–Fuchs equation of order four. This explains why we have defined an extra function in (4.5) that does not appear in (4.4). Thus, the four-dimensional space makes it very difficult to study the global behavior of the curve $\Sigma_{a,b}$, by using this Picard–Fuchs equation directly, except for some of its local properties for h near 0 and near $1/6$, shown in the next result. As all the following properties follow equivalently for $\tilde{\Sigma}_{a,b}$, we will omit the distinction of them.

Lemma 4.7. *For any $(a, b) \in G$, the curvature of $\Sigma_{a,b}$ near its two endpoints is non-zero.*

This result is equivalent to saying that for a generic quadratic Hamiltonian system the order of the Hopf bifurcation and of the homoclinic bifurcation is at most 2, and it basically follows from [11] and [153], respectively.

Taking the derivative on $I(h)$ twice, we get from (4.6) that

$$I''(h) = \alpha M''(h) + \beta X''(h) + \gamma Y''(h) = M''(h)\big[\alpha + \beta\nu(h) + \gamma\omega(h)\big],$$

where

$$\nu(h) = \frac{X''(h)}{M''(h)}, \quad \omega(h) = \frac{Y''(h)}{M''(h)}. \tag{4.8}$$

Note that $M'(h)$ is the period function of γ_h and is monotone for quadratic Hamiltonian vector fields (see [68]), hence $M''(h) \neq 0$. By our choice of h, we have $M''(h) > 0$. We define the curve in the (ν, ω)-plane

$$\Omega_{a,b} = \big\{(\nu(h), \omega(h)) : 0 \le h \le 1/6\big\}.$$

Hence the number of zeros of $I''(h)$ equals the number of intersection points (counting multiplicities) of the curve $\Omega_{a,b}$ with the straight line

$$\tilde{L}_{\alpha\beta\gamma} = \big\{(\nu, \omega) : \alpha + \beta\nu + \gamma\omega = 0\big\}.$$

Lemma 4.8. *For any $(a, b) \in G$, the following statements hold, which imply the regularity of the curve $\Omega_{a,b}$.*

(i) $\big[\nu'(h)\big]^2 + \big[\omega'(h)\big]^2 \neq 0$ *for $h \in (0, 1/6)$, and*

(ii) $(\nu(h_1), \omega(h_1)) \neq (\nu(h_2), \omega(h_2))$ *for $h_1 \neq h_2$ and $h_1, h_2 \in [0, 1/6]$.*

For proving Theorem 4.5, we suppose the contrary: for some $(a, b) \in G$ the curve $\Sigma_{a,b}$ has zero curvature at some points, and we denote by $(p(h^*), q(h^*))$ the nearest such point to the endpoint $(p(0), q(0))$. By Lemma 4.7, $h^* \in (0, 1/6)$. Now we denote the arc of $\Sigma_{a,b}$ from $h = 0$ to $h = h^*$ by $\Sigma_{a,b}^*$. We will prove the following property along this arc.

Lemma 4.9. *For any* $(a, b) \in G$ *the following statements hold.*

(i) *Along* $\Sigma_{a,b}^*$ *for* $h \in (0, h^*)$ *we have*

$$\frac{d^2q}{dp^2} > 0, \quad k_0(a, b) < \frac{dq}{dp} < k_1(a, b),$$

where

$$k_0(a, b) = \lim_{h \to 0^+} \frac{dq}{dp} = \frac{b}{a - 1} < 0, \quad k_1(a, b) = \lim_{h \to 1/6^-} \frac{dq}{dp} = \frac{q(1/6)}{p(1/6) - 1} > 0.$$

(ii) *The curve* $\Sigma_{a,b}^*$ *is smooth and* $p'(h) > 0$ *for* $h \in [0, h^*]$.

If such h^* *does not exist, then the above statements hold along* $\Sigma_{a,b}$ *for* $h \in (0, 1/6)$.

One of the crucial steps in this proof is to identify the (p, q)-plane with the (ν, ω)-plane, hence the two straight lines $L_{\alpha\beta\gamma}$ and $\tilde{L}_{\alpha\beta\gamma}$ are identified. We denote the set of tangent lines to $\Sigma_{a,b}$ (resp. $\Omega_{a,b}$) by $T_{\Sigma_{a,b}}$ (resp. $T_{\Omega_{a,b}}$), that is

$$T_{\Sigma_{a,b}} = \{\xi_h : \text{the tangent line to } \Sigma_{a,b} \text{ at } (p(h), q(h)), h \in [0, 1/6]\},$$
$$T_{\Omega_{a,b}} = \{\eta_h : \text{the tangent line to } \Omega_{a,b} \text{ at } (\nu(h), \omega(h)), h \in [0, 1/6]\}.$$

We will prove

Lemma 4.10. *For any* $(a, b) \in G$ *we have:*

(i) $\xi_t \cap \Omega_{a,b} \neq \emptyset$ *for any* $t \in (0, 1/6)$.

(ii) $\xi_0 \cap \Omega_{a,b} = \{(\nu(0), \omega(0))\}$, $\xi_{1/6} \cap \Omega_{a,b} = \{(\nu(1/6), \omega(1/6))\}$, *and the crossing is transversal.*

(iii) $\{(\nu(0), \omega(0)) \cup (\nu(1/6), \omega(1/6))\} \cap \xi_h = \emptyset$ *for any* $\xi_h \in T_{\Sigma_{a,b}^*}$ *and any* $h \in (0, h^*]$.

(iv) $\Sigma_{a,b}^*$ *and* $\Omega_{a,b}$ *have no common tangent line.*

Now, assuming Lemmas 4.7–4.10 we can give a proof of the basic Theorem 4.5.

Proof of Theorem 4.5. If the curve $\Sigma_{a,b}$ is not globally convex with non-zero curvature for all $(a, b) \in G$, then there exist a $(a^*, b^*) \in G$ and an $h^* \in [0, 1/6]$ such that the curvature of Σ_{a^*,b^*} at h^* is zero. By Lemma 4.7, $h^* \in (0, 1/6)$ and we may choose it in such a way that it is the nearest one to the endpoint $(p(0), q(0))$ with zero curvature. Lemma 4.9 implies the sectorial property and convexity (with non-zero curvature) of the curve Σ_{a^*,b^*}^* (i.e. a piece of the curve Σ_{a^*,b^*} with h restricted to $[0, h^*)$). We move $\xi_t \in T_{\Sigma_{a^*,b^*}^*}$ with the tangent point $(p(t), q(t))$ along Σ_{a^*,b^*}^* as t increases from 0 to h^*, and consider the number of intersection points of $\xi_t \cap \Omega_{a^*,b^*}$. By statements (i) and (ii) of Lemma 4.10 and Lemma 4.7,

$\xi_t \cap \Omega_{a^*,b^*}$ consists of exactly one point if $0 < t \ll 1$, counting its multiplicity. As we have supposed that $\Sigma^*_{a^*,b^*}$ has zero curvature at $h = h^* \in (0, 1/6)$, by taking $L_{\alpha\beta\gamma} = \xi_{h^*}$ the function $I(h)$ has at least a triple zero at $h = h^*$ plus a zero at $h = 0$ (Lemma 4.2.(i)), this implies that $\xi_{h^*} \cap \Omega_{a^*,b^*}$ consists of at least two points. By Lemma 4.9 and statements (ii) and (iii) of Lemma 4.10, the two endpoints of Ω_{a^*,b^*} stay on different sides of ξ_h for all $h \in (0, h^*]$. Hence, the increase in the number of intersection points of $\xi_t \cap \Omega_{a^*,b^*}$, as t increases from 0 to h^*, causes the existence of a level $\tilde{h} \in (0, h^*)$ such that $\xi_{\tilde{h}} \in T_{\Sigma^*_{a^*,b^*}}$ is also tangent to Ω_{a^*,b^*}. This contradicts Lemma 4.10.(iv).   □

Before proving the previous lemmas, we give a more geometric explanation of the above proof. By Lemma 4.2 for any α, β, and γ we have $I(0) = 0$. If we choose $L_{\alpha\beta\gamma}$ such that it is tangent to $\Sigma_{a,b}$ at a point $(p(t), q(t))$, then the graph of $I(h)$ has at least one inflection point at some value $\tilde{t} \in (0, t)$, see Figure 8.(a). In other words, in the identified (p, q) and (ν, ω) planes, the tangent line ξ_t to $\Sigma_{a,b}$

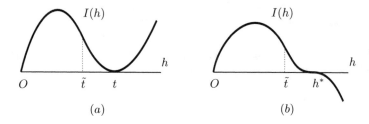

Figure 8: The behavior of curves $I(h)$.

at $(p(t), q(t))$ must cross the curve $\Omega_{a,b}$ at a point $(\nu(\tilde{t}), \omega(\tilde{t}))$ with $0 < \tilde{t} < t$, see Figure 9. If h^* is the (first) zero-curvature point on $\Sigma_{a,b}$, then h^* is at least a triple

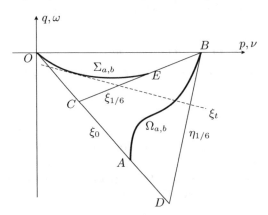

Figure 9: The relative positions of $\Sigma_{a,b}$, $\Omega_{a,b}$, and ξ_t.

zero of $I(h)$, hence the graph of $I(h)$ has at least two inflection points, one at h^* and another at some value $\tilde{t} \in (0, h^*)$, see Figure 8.(b). This means that the tangent line $\xi_t = L_{\alpha\beta\gamma}$ of $\Sigma_{a,b}$ crosses the curve $\Omega_{a,b}$ at least at two points. Lemma 4.10 tells us that the tangent line ξ_0 (to $\Sigma_{a,b}$ at the endpoint $(p(0), q(0)) = (0, 0)$) crosses $\Omega_{a,b}$ at its (left) endpoint A while the tangent line $\xi_{1/6}$ (to $\Sigma_{a,b}$ at the endpoint $E = (p(1/6), q(1/6))$) crosses $\Omega_{a,b}$ at its (right) endpoint B, and when t moves from 0 to 1/6, the two points A and B always stay (respectively) below and above the tangent line ξ_t. Hence the increasing of the number of intersection points of ξ_t with $\Omega_{a,b}$ must be through a position that ξ_t is also tangent to $\Omega_{a,b}$. But Lemma 4.10.(iv) excludes this possibility, hence the transversality of ξ_t with Ω for $0 \le t \ll 1$ remains true for all $t \in [0, 1/6]$, and this implies the non-zero curvature along $\Sigma_{a,b}$ for all $(a, b) \in G$.

4.3 The Picard–Fuchs Equation and the Riccati Equation

When $(a, b) \in G\backslash l_\infty$, by using a standard method, such as the one described in Section 3.1, one can obtain the following Picard–Fuchs equation of order 4, satisfied by $X(h)$, $Y(h)$, $M(h)$, and $K(h)$ (see Lemma 3.3 of [154]):

$$
\begin{aligned}
&- 6bhM' + bX' - (a+1)Y' - 2a(a+1)K' + 4bM = 0, \\
&(6\lambda h + a + 1)Y' + (8a^2 + b^2 + 4a)K' + b(a+1)M - 6\lambda Y = 0, \\
&b\lambda(6h - 1)X' + a(\lambda - 2a(a+1))Y' + ((4a^2 + 3a + 1)\lambda \\
&\quad - 8a^3(a+1)^2)K' - 6b\lambda X + b(\lambda - 2a^2(a+1))M = 0, \\
&((-6h\lambda^2 + b^4 + (12a^2 + 5a + 1)b^2 + 4a^3(7a + 3))K' + a(8a^2 + b^2 + 4a) \\
&\cdot (Y' + bM) + \lambda(-(a+1)bX + 8\lambda K + (4a^2 + b^2)Y) = 0,
\end{aligned}
\tag{4.9}
$$

where $' = \frac{d}{dh}$, and $\lambda = 4a^3 - b^2$. Note that $\lambda = 0$ corresponds to $(a, b) \in l_\infty$, and in this case the last three equations in (4.9) are not independent, and a Picard–Fuchs equation of order 3, similar to (4.9), can be obtained ($K(h)$ and $K'(h)$ do not appear). Hence the results, parallel to Lemma 4.11 and equation (4.14) in this section, can be obtained for this special value $\lambda = 0$. In fact, they are limits as $\lambda \to 0$ of the corresponding results here. Hence the discussions are valid for all $(a, b) \in G$.

For simplicity, we use the notation

$$
\begin{aligned}
\lambda_1(a, b) &= (1 - a)^2(2a + 1) - b^2, \\
\lambda_2(a, b) &= (3a + 1)^2 + b^2, \\
\lambda_3(a, b) &= (3a - 1)^2 + 5b^2 + 4.
\end{aligned}
\tag{4.10}
$$

Note that $\lambda_1(a, b) > 0$ for $(a, b) \in G$. The following result follows from (4.9) (see the proof of Lemma 2 of [78] or Lemma 3.3 of [154]).

Lemma 4.11. For $0 < h \ll 1$ we have

$$p(h) = \frac{1}{2}(1-a)h - \frac{1}{72}[5(11a+1)b^2 + (a-1)(63a^2 + 18a + 55)]h^2 + O(h^3),$$

$$q(h) = -\frac{b}{2}h - \frac{b}{72}(55b^2 + 183a^2 - 42a - 5)h^2 + O(h^3).$$

Lemma 4.12. For $(a,b) \in G$ we have:

(i) $\displaystyle \lim_{h \to 0+} \frac{dq}{dp} = \frac{b}{a-1} < 0,$

(ii) $\displaystyle \lim_{h \to 0+} \frac{d^2q}{dp^2} = \frac{20}{3} \frac{b\lambda_1(a,b)}{(1-a)^3} > 0,$ and

(iii) $\displaystyle \lim_{h \to 1/6-} \frac{dq}{dp} = -\frac{q(1/6)}{1-p(1/6)} > 0.$

Proof. Statements (i) and (ii) are easily deduced from Lemma 4.11. Statement (iii) can be proved by using the expansions of $M(h)$, $X(h)$, and $Y(h)$ in h near $1/6$ as follows,

$$c_1 + c_2(h - 1/6)\ln(1/6 - h) + c_3(h - 1/6) + o(h - 1/6),$$

see [291] or (1.8) of [154]. □

Taking derivatives with respect to h in the first three equations of (4.9), and removing M', we can express X'', K'' through M'', Y'' as follows,

$$\begin{aligned} X'' &= d_1(h)M'' + d_2(h)Y'', \\ K'' &= d_3(h)M'' + d_4(h)Y'', \end{aligned} \tag{4.11}$$

where

$$d_1(h) = \frac{6\lambda_1(a,b)h}{L(h)},$$

$$d_2(h) = \frac{[12(3a^2 + 2a + 1)b^2 - 24a^3(3a+1)(a-1)]h + (a-1)\lambda_2(a,b)}{bL(h)},$$

$$d_3(h) = \frac{-6b(a+1)(6h-1)h}{L(h)}, \tag{4.12}$$

$$d_4(h) = \frac{6[12(4a^3 - b^2)h + b^2 - 6a^3 - 3a^2 + 1]h}{L(h)},$$

$$L(h) = 12(a^3 - 6a^2 - 3a - b^2)h + \lambda_2(a,b).$$

Note that $L(0) > 0$ and $L(1/6) = \lambda_1(a,b) > 0$ (see (4.10)), hence the linear function $L(h) \neq 0$ for all $h \in [0, 1/6]$.

Taking derivatives in (4.9) with respect to h once more, and using (4.11) we get

$$T(h)\frac{d}{dh}\begin{pmatrix} M'' \\ Y'' \end{pmatrix} = \begin{pmatrix} e_1(h) & e_2(h) \\ e_3(h) & e_4(h) \end{pmatrix}\begin{pmatrix} M'' \\ Y'' \end{pmatrix}, \tag{4.13}$$

where

$$T(h) = -6bh(6h-1)L(h)\tilde{T}(h),$$

$$\tilde{T}(h) = 36(4a^3 - b^2)^2 h^2 - 6[b^4 + 2(6a^2 + 3a + 1)b^2 + 8a^3(3a+1)]h + \lambda_2(a,b),$$

and

$$e_i(h) = \sum_{k=0}^{4} e_{ik}h^k,$$

with e_{ik} polynomials in a and b. We omit their expressions here; our readers can find them in [32].

From the definition of $\omega(h)$ in (4.8), we have

$$\omega'(h) = \frac{(Y''(h))'}{M''(h)} - \frac{(M''(h))'}{M''(h)}\omega(h).$$

Combining this fact with (4.13), we obtain a two-dimensional system of equations

$$\dot{h} = T(h), \qquad \dot{\omega} = \phi(h,\omega), \tag{4.14}$$

where $\phi(h,\omega) = -e_2(h)\omega^2 + (e_4(h) - e_1(h))\omega + e_3(h)$, and the dot denotes the derivative with respect to an arbitrary variable s. Note that system (4.14) is equivalent to a Riccati equation with dependent variable ω and independent h.

Remark 4.13. We note that $T(h) \neq 0$ for $h \in (0, 1/6)$, hence system (4.14) has no singularities for $h \in (0, 1/6)$. In fact, we have shown that $L(h)$ has no zeros for $h \in (0, 1/6)$. If $(a,b) \in G_1$, then $\tilde{T}(h)$ has no real roots. If $(a,b) \in G_2 \cup G_3 \cup l_2 \cup l_\infty$, then the roots of $\tilde{T}(h)$ correspond to other singularities of X_H, besides the center $O(0,0)$ and the saddle $S(1,0)$. By the monotonic property of the level curves of the Hamiltonian vector field and the relative positions of the singularities, we immediately obtain that the roots of $\tilde{T}(h)$ must be greater than $1/6$.

By Remark 4.13 and direct computations we obtain the following result.

Lemma 4.14. *For $h \in [0, 1/6]$ system (4.14) has four singularities: two improper nodes at $(0,0)$ and $(1/6, 0)$ and two hyperbolic saddles at $(0, \omega_0)$ and $(1/6, \omega_1)$, where*

$$\omega_0 = \frac{-6b}{\lambda_3(a,b)} < 0, \qquad \omega_1 = \frac{-6b(2a+1)}{5b^2 - 82a^3 - 93a^2 - 36a - 5}.$$

When the denominator $5b^2 - 82a^3 - 93a^2 - 36a - 5$ goes to zero, the singularity $(1/6, \omega_1)$ goes to infinity.

We recall that an *improper node* is a node such that all the orbits arrive to or exit from it in one direction.

Let

$$C_\omega = \{(h, \omega) : 0 \le h \le 1/6, \ \omega = \omega(h) \text{ is defined in (4.8)}\}.$$

The following lemma can be proved in the same way as Lemma 3.1 in [199], except statement (ii) which is a consequence of Lemma 4.21 below. Statement (i) of the next lemma shows that C_ω is the unstable manifold from the saddle $(0, \omega_0)$ to the improper node $(1/6, 0)$ of system (4.14).

Lemma 4.15.   (i) $\displaystyle\lim_{h \to 0+} \omega(h) = \omega_0$, $\displaystyle\lim_{h \to 1/6-} \omega(h) = 0$.

(ii) $\omega(h) < 0$ for $h \in (0, 1/6)$.

(iii) $\displaystyle\lim_{h \to 0+} \nu(h) = \nu_0$, $\displaystyle\lim_{h \to 1/6-} \nu(h) = 1$, where

$$\nu_0 = \frac{6(1 - a)}{\lambda_3(a, b)} > 0.$$

(iv) *We have*

$$\lim_{h \to 0+} \omega'(h) = \frac{5}{2} \frac{b f_1(a, b)}{(\lambda_3(a, b))^2}, \qquad \lim_{h \to 0+} \nu'(h) = \frac{5}{2} \frac{f_2(a, b)}{(\lambda_3(a, b))^2},$$

$$\lim_{h \to 0+} [\omega''(h)\nu'(h) - \omega'(h)\nu''(h)] = \frac{175}{6} \frac{b\lambda_1(a, b)\lambda_2(a, b) f_3(a, b)}{(\lambda_3(a, b))^3},$$

where

$$f_1(a, b) = 7b^4 + (42a^2 + 60a - 70)b^2 - 189a^4 + 180a^3 - 174a^2 - 12a + 67,$$
$$f_2(a, b) = (7a - 67)b^4 + (162a^3 - 270a^2 - 58a + 70)b^2 + (a - 1)(3a + 1)$$
$$\cdot (9a^3 - 27a^2 - 21a + 7),$$
$$f_3(a, b) = 55b^4 + (126a^2 - 204a - 106)b^2 - 81a^4 + 324a^3 + 162a^2 - 204a + 55.$$

(v) *We have*

$$\lim_{h \to 1/6-} \frac{\nu'(h)}{\omega'(h)} = -\frac{(2a + 1)(3a + 1)}{b}.$$

$\square$

From (4.8) and (4.11) we obtain the expression of $\nu(h)$ as a function of h and $\omega(h)$ as follows

$$\nu(h) = d_1(h) + d_2(h)\omega(h), \tag{4.15}$$

where $d_i(h) = d_i(h; a, b)$, for $i = 1, 2$ are given in (4.12).

We consider the following transformation from the (h, ω)-plane to the (ν, ω)-plane:

$$\nu = d_1(h) + d_2(h)\omega, \qquad \omega = \omega. \tag{4.16}$$

It is easy to see that (4.16) maps the straight line $\{(h, \omega) : h = h_0\}$ ($h_0 \in [0, 1/6]$) in the (h, ω)-plane to a straight line in the (ν, ω)-plane. In particular, it maps $\{(h, \omega) : h = 0\}$ to L_0, and $\{(h, \omega) : h = 1/6\}$ to L_3, being

$$L_0 = \left\{(\nu, \omega) : \nu = \frac{a-1}{b}\omega\right\}, \ L_3 = \left\{(\nu, \omega) : \nu = 1 - \frac{(2a+1)(3a+1)}{b}\omega\right\}. \tag{4.17}$$

We note that if $a = 0$, then L_0 is parallel to L_3, and if $a \neq 0$, then $L_0 \cap L_3 = \{(\tilde{\nu}, \tilde{\omega})\}$, where

$$\tilde{\nu} = \frac{a-1}{6a(a+1)}, \quad \tilde{\omega} = \frac{b}{6a(a+1)}. \tag{4.18}$$

Let

$$D_{a,b} = \{(h, \omega) \in \mathbb{R}^2 : 0 \leq h \leq 1/6; \ -\infty < \omega < \infty \text{ if } a = 0,$$
$$-\infty < \omega < \tilde{\omega} \text{ if } a > 0, \ \tilde{\omega} < \omega < \infty \text{ if } a < 0\}.$$

Correspondingly, let $\tilde{D}_{a,b}$ be the region in the (ν, ω)–plane, which is the strip $-\frac{1}{b}\omega \leq \nu \leq -\frac{1}{b}\omega + 1$ if $a = 0$; and is the corresponding sector region, limited by the two straight lines L_0 and L_3 with vertex at $(\tilde{\nu}, \tilde{\omega})$ if $a \neq 0$, see Figure 10. Note that the vertex $(\tilde{\nu}, \tilde{\omega})$ is not included in $\tilde{D}_{a,b}$. The Jacobian of transformation (4.16),

$$\frac{D(\nu, \omega)}{D(h, \omega)} = d_1'(h) + d_2'(h)\omega = -\frac{6\lambda_1(a,b)\lambda_2(a,b)[6a(a+1)\omega - b]}{bL^2(h)},$$

is non-zero if $a = 0$, and is zero only for $\omega = \tilde{\omega}$ if $a \neq 0$. Hence, we immediately have the following result.

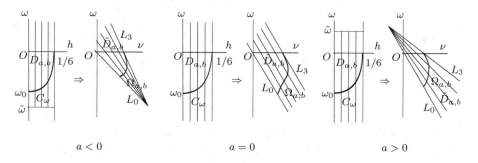

Figure 10: From $D_{a,b}$ to $\tilde{D}_{a,b}$ through the transformation (4.16).

Lemma 4.16. *For any $(a,b) \in G$ the transformation (4.16) from $D_{a,b}$ to $\tilde{D}_{a,b}$ is a smooth diffeomorphism. Hence, system (4.14) in $D_{a,b}$ becomes the smooth system*

$$\dot{\nu} = \varphi_1(\nu, \omega), \qquad \dot{\omega} = \varphi_2(\nu, \omega), \tag{4.19}$$

in $\tilde{D}_{a,b}$.

From Remark 4.13 and Lemmas 4.14 and 4.16 we obtain the following result.

Lemma 4.17. *For $(a.b) \in G$ we have*

(i) *Any orbit of system (4.14), especially C_ω, is transversal to all parallel lines $\{h = h_0, h_0 \in [0, 1/6)\}$ in $D_{a,b}$.*

(ii) *Any orbit of system (4.19), especially $\Omega_{a,b}$, is transversal to all straight lines between L_0 and L_3 in $\tilde{D}_{a,b}$, the lines are parallel if $a = 0$, or are in the sectorial region with vertex $(\tilde{\nu}, \tilde{\omega})$ if $a \neq 0$, see Figure 10.*

We denote by $L^*_{\alpha\beta\gamma}$ the part of the straight line $\tilde{L}_{\alpha\beta\gamma}$ in the (ν, ω)-plane, which is contained in $\tilde{D}_{a,b}$. Let $C_U = \{(h, \omega) : 0 \le h \le 1/6, \ \omega = U(h)\}$ where

$$U(h) = U(h; a, b, \alpha, \beta, \gamma) = \frac{z_1 h + z_0}{n_1 h + n_0} =: \frac{Z(h)}{N(h)}, \tag{4.20}$$

with

$$z_1 = 6b[2(b^2 - a^3 + 6a^2 + 3a)\alpha - \lambda_1(a, b)\beta],$$
$$z_0 = -ba\lambda_2(a, b),$$
$$n_1 = 12b(a^3 - 6a^2 - 3a - b^2)\gamma + [12(3a^2 + 2a + 1)b^2 - 24a^3(3a + 1)(a - 1)]\beta,$$
$$n_0 = \lambda_2(a, b)[(a - 1)\beta + b\gamma],$$

and λ_i defined in (4.10).

Lemma 4.18. *For any $(a, b) \in G$ and any constants α, β, and γ, the straight line $L^*_{\alpha\beta\gamma}$ is tangent to an orbit of system (4.19) of order k (in particular to $\Omega_{a,b}$, at a point $(\nu(h_0), \omega(h_0))$ for $h_0 \in (0, 1/6)$), if and only if C_U is tangent to the corresponding orbit of system (4.14) of order k (in particular to C_ω, at $(h_0, \omega(h_0)))$.*

Proof. Under the transformation (4.16), the line $\tilde{L}_{\alpha\beta\gamma}$ becomes

$$\alpha + \beta\nu + \gamma\omega = \frac{N(h)\omega - Z(h)}{bL(h)} = 0, \tag{4.21}$$

where $L(h) \neq 0$ for $h \in [0, 1/6]$ is given in (4.12), and the linear functions $N(h)$ and $Z(h)$ are defined in (4.20). If $N(h_0) \neq 0$, then for h near h_0 we can rewrite the above equality as

$$\alpha + \beta\nu + \gamma\omega = \frac{N(h)}{bL(h)}[\omega - U(h)] = 0.$$

This means that the transformation (4.16) maps the straight line $L^*_{\alpha\beta\gamma}$ to the curve C_U, and the lemma is proved for h near h_0 by Lemma 4.16. Next, we show that we can skip all zero points of $N(h)$ for $h \in [0, 1/6]$. In fact, if $N(h_0) = 0$ but $Z(h_0) \neq 0$, then equation (4.21) is not satisfied and we do not need to consider it. If $N(h_0) = Z(h_0) = 0$, then the resultant of $N(h)$ and $Z(h)$ must be zero. By a direct computation and using $(a, b) \in G$, we obtain

$$\beta \left[6a(a+1)\alpha + (a-1)\beta + b\gamma \right] = 0.$$

If $\beta = 0$, then $N(h) = bL(h)\gamma$ and $Z(h) = -bL(h)\alpha$, which contradicts the non-zero property of $L(h)$ for $h \in [0, 1/6]$. If $6a(a+1)\alpha + (a-1)\beta + b\gamma = 0$, then $L^*_{\alpha\beta\gamma} \in \tilde{D}_{a,b}$ is parallel to L_0 and L_3 when $a = 0$, or passes through the vertex $(\tilde{\nu}, \tilde{\omega})$ of the sector when $a \neq 0$, see (4.17) and (4.18). By Lemma 4.17, there is no orbit of system (4.19) tangent to it, and the assumption of the lemma is not satisfied. □

Lemma 4.19. *For any $(a, b) \in G$ and any constants α, β, and γ, there exist at most four points on $L^*_{\alpha\beta\gamma}$, counting their multiplicities, such that at each of these points the vector field (4.19) is tangent to $L^*_{\alpha\beta\gamma}$. In particular, if one of the endpoints of $L^*_{\alpha\beta\gamma}$ is $(\nu(0), \omega(0))$ or $(\nu(1/6), \omega(1/6))$, then the endpoint is included in these tangent points.*

Proof. By Lemma 4.18 we only need to consider the number of tangent points on C_U (corresponding to $L^*_{\alpha\beta\gamma}$) with respect to the vector field (4.14) in the (h, ω)-plane. By using (4.14) and (4.20) we obtain

$$\dot{\omega} - U'(h)\dot{h}\big|_{\omega=U(h)} = \phi(h, U(h)) - U'(h)T(h) = \frac{b^2 L^2(h)\, F(h)}{N^2(h)}, \tag{4.22}$$

where $F(h) = F(h; a, b, \alpha, \beta, \gamma)$ is a polynomial in all its arguments, and of degree 4 in h. Besides, $F(h)$ has the factor h or $(h - 1/6)$ if $L^*_{\alpha\beta\gamma}$ has the endpoint $(\nu(0), \omega(0))$ or $(\nu(1/6), \omega(1/6))$, respectively. Note that we may suppose that $N(h) \neq 0$ for $h \in [0, 1/6]$, see the proof of Lemma 4.18. □

By using the variation of argument and the fact that the function $F(h)$ in (4.22) is a polynomial in h of degree 4, we can prove the following results.

Lemma 4.20. *For any $(a, b) \in G$, if $\Omega_{a,b}$ has an inflection point, then the tangent line to $\Omega_{a,b}$ at this point does not pass through the point $C = \xi_0 \cap \xi_{1/6}$ (see Figure 9).*

Lemma 4.21. *For any $(a, b) \in G$:*

(i) *The curve $\Omega_{a,b}$ is located in the region $\tilde{D}_{a,b}$, i.e. between the lines ξ_0 and $\eta_{1/6}$, and on the right-hand side of the line $\xi_{1/6}$, except the endpoint $B = (\nu(1/6), \omega(1/6)) = (1, 0)$.*

(ii) *The curve $\Sigma_{a,b}$ is located inside the closed triangle with vertices $O, C,$ and B, denoted by $\Delta_{a,b}$, see Figure 9.*

4.4   Outline of the Proofs of the Basic Lemmas

Below we give an outline of the proofs of Lemmas 4.8–4.10, see [32] for detailed proofs.

Proof of Lemma 4.8. From (4.15) we see that the transformation (4.16) maps C_ω to $\Omega_{a,b}$, and by Lemmas 4.14 and 4.15, C_ω, satisfying $\dot h \neq 0$ for $h \in (0, 1/6)$, is a regular curve. Hence, by Lemma 4.16, to prove the regularity of $\Omega_{a,b}$ it is enough to show that C_ω stays in $D_{a,b}$, i.e. C_ω does not meet the straight line $\{\omega = \tilde\omega\}$ in the (h, ω)-plane for $a \neq 0$. This fact can be proved basically by using the two-dimensional system (4.14) and the equality (4.22), which gives the number of contact points of system (4.14) with the curve $\omega = U(h)$.   □

Proof of Lemma 4.9. (i) Lemma 4.21 shows that for any $(a, b) \in G$ the centroid curve $\Sigma_{a,b}$ is located inside the triangle region $\Delta_{a,b}$ (see Figure 9) and the curve $\Omega_{a,b}$ is located on the right-hand side of the straight line BC (i.e. $\xi_{1/6}$). From Lemma 4.12 we see that $d^2q/dp^2 > 0$ and dq/dp at $(p(h), q(h)) \in \Sigma_{a,b}$ is increasing from $k_0(a, b)$ as h increases from 0, until h^* or the first value h_∞ that $\lim_{h\to h_\infty} dq/dp = \infty$. We claim that the latter case is impossible. In fact, $dq/dp < k_1(a, b)$ for any point on $\Sigma_{a,b}$. If this is not true, then we would find a point on $\Sigma_{a,b}$, such that the tangent line at this point is parallel to $\xi_{1/6}$, hence it has no intersection with the curve $\Omega_{a,b}$, giving a contradiction, since any tangent line to $\Sigma_{a,b}$ must cross $\Omega_{a,b}$.

(ii) By using the fact that any tangent line to $\Omega_{a,b}$ does not pass through the point $D(\tilde\nu, \tilde\omega)$ (see Figure 10 and (4.18)) one can prove that $p(h) < \nu(h)$ for $h \in (0, h^*)$, and this implies $p'(h) > 0$ for $h \in (0, h^*)$. In fact, Lemma 4.11 implies $p(0) = 0$ and $p'(0) > 0$ for $(a, b) \in G$. We suppose that $h_0 = \inf_{h\in(0,h^*)}\{h : p'(h) = 0\}$, then $h_0 > 0$, $p'(h_0) = 0$, and $p'(h) > 0$ for $0 < h_0 - h \ll 1$. By (4.7) and (4.8), $X'(h_0)M(h_0) - M'(h_0)X(h_0) = 0$, and for $0 < h_0 - h \ll 1$ we have

$$\begin{aligned}
M^2(h)p'(h) &= (X'(h)M(h) - M'(h)X(h)) - (X'(h_0)M(h_0) - M'(h_0)X(h_0)) \\
&= (X''(\theta)M(\theta) - M''(\theta)X(\theta))(h - h_0) \\
&= M(\theta)M''(\theta)[\nu(\theta) - p(\theta)](h - h_0) < 0,
\end{aligned}$$

when $\theta \in (h, h_0)$, since for all $h \in (0, h^*)$ we have $M(h) > 0$ (the area of γ_h), $M''(h) > 0$ (the derivative of the period function which is positive by [68]), and $\nu(\theta) - p(\theta) > 0$. This gives a contradiction.   □

Proof of Lemma 4.10. Statement (i) was explained below the proof of Theorem 4.5: it can be seen clearly from Figure 8.(a) that for any double zero t $(t > 0)$ of $I(h)$ there exists at least one inflection point of $I(h)$ at some point $\tilde t \in (0, t)$. Statement (ii) follows from statement (i) and Lemma 4.12.(iii), and the transversality can be obtained by direct computation. To prove statement (iii), we note that $\Sigma^*_{a,b}$ is convex, hence the fact $(\nu(0), \omega(0)) \cap \xi_h = \emptyset$ for $h \in (0, h^*]$ is obviously true. If $(\nu(1/6), \omega(1/6)) \cap \xi_h \neq \emptyset$ for some $h \in (0, h^*]$, then, by the facts that $\Sigma_{a,b}$

is located inside the triangle region OCB (see Figure 9) and the curve $\Omega_{a,b}$ is located on the right-hand side of the straight line $\xi_{1/6}$, we conclude that the point $B = (\nu(1/6), \omega(1/6))$ is the only intersection point of $\Omega_{a,b} \cap \xi_h$, and this contradicts the fact that ξ_h must intersect $\Omega_{a,b}$ at some point $(\nu(t), \omega(t))$ for $t \in (0, h)$.

Finally, we prove statement (iv): there is no common tangent line for the two curves $\Sigma_{a,b}$ and $\Omega_{a,b}$. The idea is to consider the motion of the tangent line η_h, with tangent point moving on $\Omega_{a,b}$, as h decreases from $1/6$. Since $\Omega_{a,b}$ is on the right-hand side of the line BC (see Figure 9), there are only two possibilities for η_h also tangent to $\Sigma_{a,b}$: either η_h passes over the point C upwards, and enters the triangle region OBC, or η_h passes the points $B(1,0)$, $O(0,0)$, and over $\Sigma_{a,b}$ downwards, and gets a tangent position. The latter case contradicts statement (ii) of Lemma 4.17. In the former case, by using a deformation argument we would find a $(\tilde{a}, \tilde{b}) \in G$, such that $\Omega_{\tilde{a}, \tilde{b}}$ has an inflection point, and its tangent line at this point passes through the point C, contradicting Lemma 4.20. $\qquad\square$

4.5 Proof of Theorem 4.6

For $(a, b) \in G_2$, the Hamiltonian vector field X_H has two centers C, $\hat{C}$, two saddles S, $\hat{S}$, two saddle loops $\mathcal{L}$, $\hat{\mathcal{L}}$, and the corresponding period annuli $D(\mathcal{L})$, $D(\hat{\mathcal{L}})$. Hence we have two centroid curves $\Sigma \subset D(\mathcal{L})$ and $\hat{\Sigma} \subset D(\hat{\mathcal{L}})$. For the sake of simplicity in the exposition we do not add the subscript (a, b) as we have used in all the section to denote the dependence on the parameters. Since the convexity does not change under affine transformations, we can move C or $\hat{C}$ (resp. S or $\hat{S}$) to $(0,0)$ (resp. $(1,0)$) and obtain the normal form (4.1) by an affine transformation, so from Theorem 4.5 we conclude that both Σ and $\hat{\Sigma}$ are strictly convex. Note that as X_H is a quadratic system, the four singularities form a quadrilateral with C and $\hat{C}$ as a pair of opposite vertices and S and $\hat{S}$ as another opposite pair (see Remark 4.22). If we exchange C to $\hat{C}$ and S to $\hat{S}$ by doing an affine transformation, then we must reverse the direction of one coordinate axis (or with a rotation π), hence Σ and $\hat{\Sigma}$ must be one convex and the other concave.

Now we denote by L_c (resp. L_s) the straight line passing through C and $\hat{C}$ (resp. S and $\hat{S}$); by O the intersection point of L_c and L_s; by Δ (resp. $\hat{\Delta}$) the interior of the triangle with vertices C, S, and O (resp. $\hat{C}$, $\hat{S}$, and O). Next we denote by t_c the straight half-line which is tangent to Σ at C and points to the direction of the convexity; by t_s the straight half-line from S to another endpoint Z of Σ (the centroid point of $D(\mathcal{L})$); by τ the intersection point of t_c and t_s. By Lemma 4.10.(ii), t_s is tangent to Σ at Z (note that the point $(\nu(1/6), \omega(1/6))$ corresponds to a saddle), and by Lemma 4.9, τ is located on the same side of the convexity of Σ. We similarly define the straight half-lines $\hat{t}_c$, $\hat{t}_s$, and let $\hat{\tau} = \hat{t}_c \cap \hat{t}_s$, see Figure 11.

As has been pointed out by E. Horozov & I.D. Iliev [152] to finish the proof of Theorem 4.6, it is enough to show that $\tau \in \Delta$ and $\hat{\tau} \in \hat{\Delta}$ for any $(a, b) \in G_2$. This follows from the claims:

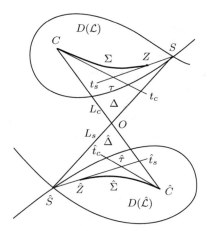

Figure 11: The relative positions of the two centroid curves.

(1) t_s and $\hat{t}_s$ are located on different sides of L_s; and

(2) t_c and $\hat{t}_c$ are located on different sides of L_c.

Claim (1) follows from the simple fact that for a quadratic system on any straight line there are at most two points at which the vector field is tangent to this line ([67, 326]). Now L_s passes through the two singularities S and $\hat{S}$, so it must stay outside $D(\mathcal{L})$ and $D(\hat{\mathcal{L}})$. Otherwise, one orbit inside $D(\mathcal{L})$ or $D(\hat{\mathcal{L}})$ would be tangent to it. On the other hand, the point Z (resp. $\hat{Z}$) is inside $D(\mathcal{L})$ (resp. $D(\hat{\mathcal{L}})$).

Claim (2) can be verified by means of a direct calculation. For the normal form (4.1), $C(0,0)$ is a center and the slope of t_c is $k_0 = b/(a-1)$. Note that the equations of l_2 and l_∞ are given by $b = \sqrt{-4a(2a+1)}, -1/2 < a < 0$ and $b = 2\sqrt{a^3}, 0 < a < 1/2$, respectively. We find that the conditions for $(a,b) \in G_2$ is: $\xi_1 := b^2 - 4a^3 > 0$, $\xi_2 := b^2 + 4a(2a+1) > 0$, and $\xi_3 := (1-a)^2(2a+1) - b^2 > 0$, where $|a| < 1/2$ and $0 < b < 1$. The other center is $\hat{C}(\tilde{x}, \tilde{y})$ with

$$\tilde{x} = \frac{4a^2 + b^2 + b\sqrt{\xi_2}}{2\xi_1}, \quad \tilde{y} = \frac{-(2a\tilde{x}+1)}{b}.$$

Hence, the slope of L_c is $\tilde{k} = \tilde{y}/\tilde{x}$, and we have

$$k_0 - \tilde{k} = \frac{\eta_1 + \eta_2\sqrt{\xi_2}}{(1-a)(4a^2 + b^2 + b\sqrt{\xi_2})},$$

where $\eta_1 = b(2 - 6a^2 - b^2)$ and $\eta_2 = 2a - 2a^2 - b^2$. From $\xi_3 > 0$ we have $-b^2 > 3a^2 - 2a^3 - 1$, which implies $\eta_1 > b(1 - 2a)(a+1)^2 > 0$ because $|a| < 1/2$

for $(a, b) \in G_2$. Hence, if $\eta_2 \geq 0$, then we have $k_0 - \tilde{k} > 0$. If $\eta_2 < 0$, then a computation gives

$$\eta_1^2 - \eta_2^2 \xi_2 = 4\xi_1 \xi_3 > 0,$$

and we obtain the same conclusion. Since we may change $\hat{t}_c$ by t_c through an affine transformation, and reverse one coordinate axis as explained before, $\hat{t}_c$ must stay on the other side of L_c.

Remark 4.22. In the proof the geometric position of the four singularities of X_H, which are hyperbolic, is used. Two of them are saddles and the other two are centers. Berlinskiĭ's Theorem ([12]) ensures that in this case they always form a quadrilateral, with the saddles in two opposite vertices an the centers in the other two. This result also ensures that the other only possibility is that three of them are in the vertices of a triangle and the fourth in its interior. This property also appears in [355]. For a simpler proof using Euler–Jacobi formula see [59].

Bibliography

[1] M. J. Álvarez, B. Coll, P. De Maesschalck, and R. Prohens. Asymptotic lower bounds on Hilbert numbers using canard cycles. *J. Differential Equations*, 268(7):3370–3391, 2020.

[2] Y. An and M. Han. On the number of limit cycles near a homoclinic loop with a nilpotent singular point. *J. Differential Equations*, 258(9):3194–3247, 2015.

[3] A. A. Andronov, E. A. Leontovich, I. I. Gordon, and A. G. Maĭer. *Qualitative theory of second-order dynamic systems*. Halsted Press (A division of John Wiley & Sons), New York-Toronto, Ont.; Israel Program for Scientific Translations, Jerusalem-London, 1973. Translated from the Russian by D. Louvish.

[4] A. A. Andronov, E. A. Leontovich, I. I. Gordon, and A. G. Maĭer. *Theory of bifurcations of dynamic systems on a plane*. Halsted Press [A division of John Wiley & Sons], New York-Toronto, Ont.; Israel Program for Scientific Translations, Jerusalem-London, 1973. Translated from the Russian.

[5] V. I. Arnol'd. Loss of stability of self-induced oscillations near resonance, and versal deformations of equivariant vector fields. *Funkcional. Anal. i Priložen.*, 11(2):1–10, 95, 1977.

[6] V. I. Arnol'd. Ten problems. In *Theory of singularities and its applications*, volume 1 of *Adv. Soviet Math.*, pages 1–8. Amer. Math. Soc., Providence, RI, 1990.

[7] V. I. Arnol'd, S. M. Guseĭn-Zade, and A. N. Varchenko. *Singularities of differentiable maps. Vol. II*, volume 83 of *Monographs in Mathematics*. Birkhäuser Boston, Inc., Boston, MA, 1988. Monodromy and asymptotics of integrals, Translated from the Russian by Hugh Porteous, Translation revised by the authors and James Montaldi.

[8] J. C. Artés, J. Llibre, and D. Schlomiuk. The geometry of quadratic differential systems with a weak focus of second order. *Internat. J. Bifur. Chaos Appl. Sci. Engrg.*, 16(11):3127–3194, 2006.

© The Author(s), under exclusive license to Springer Nature Switzerland AG 2024
C. Christopher et al., *Limit Cycles of Differential Equations*, Advanced Courses in Mathematics - CRM Barcelona, https://doi.org/10.1007/978-3-030-59656-9

[9] J. C. Artés, J. Llibre, D. Schlomiuk, and N. Vulpe. Global topological config-urations of singularities for the whole family of quadratic differential systems. *Qual. Theory Dyn. Syst.*, 19(1):Paper No. 51, 32, 2020.

[10] R. Bamón. Solution of Dulac's problem for quadratic vector fields. *An. Acad. Brasil. Ciênc.*, 57(3):265–266, 1985.

[11] N. N. Bautin. On the number of limit cycles appearing with variation of the coefficients from an equilibrium state of the type of a focus or a center. *Mat. Sbornik N.S.*, 30(72):181–196, 1952. (In Russian).

[12] A. N. Berlinskiĭ. On the behavior of the integral curves of a differential equation. *Izv. Vysš. Učebn. Zaved. Matematika*, 1960(2 (15)):3–18, 1960.

[13] G. Binyamini, D. Novikov, and S. Yakovenko. On the number of zeros of Abelian integrals. *Invent. Math.*, 181(2):227–289, 2010.

[14] T. R. Blows and N. G. Lloyd. The number of limit cycles of certain polyno-mial differential equations. *Proc. Roy. Soc. Edinburgh Sect. A*, 98(3-4):215–239, 1984.

[15] R. I. Bogdanov. Bifurcations of a limit cycle of a certain family of vector fields on the plane. *Trudy Sem. Petrovsk.*, Vyp. 2:23–35, 1976.

[16] Y. L. Bondar' and A. P. Sadovskiĭ. On a theorem of Żołądek. *Differ. Uravn.*, 44(2):263–265, 287, 2008.

[17] L. P. Bonorino, E. H. M. Brietzke, J. a. P. Lukaszczyk, and C. A. Taschetto. Properties of the period function for some Hamiltonian systems and homo-geneous solutions of a semilinear elliptic equation. *J. Differential Equations*, 214(1):156–175, 2005.

[18] M. Boutat. L'étude à l'infini de la bifurcation de Takens-Bogdanov. *C. R. Acad. Sci. Paris Sér. I Math.*, 316(2):183–186, 1993.

[19] A. Buică. On the equivalence of the Melnikov functions method and the averaging method. *Qual. Theory Dyn. Syst.*, 16(3):547–560, 2017.

[20] A. Buică, A. Gasull, and J. Yang. The third order Melnikov function of a quadratic center under quadratic perturbations. *J. Math. Anal. Appl.*, 331(1):443–454, 2007.

[21] A. Buică, J. Giné, and J. Llibre. Bifurcation of limit cycles from a polynomial degenerate center. *Adv. Nonlinear Stud.*, 10(3):597–609, 2010.

[22] A. Buică and J. Llibre. Averaging methods for finding periodic orbits via Brouwer degree. *Bull. Sci. Math.*, 128(1):7–22, 2004.

[23] A. Buică and J. Llibre. Limit cycles of a perturbed cubic polynomial differ-ential center. *Chaos Solitons Fractals*, 32(3):1059–1069, 2007.

[24] S. Cai. Recent developments in the study of quadratic systems. *Adv. in Math. (Beijing)*, 18(1):5–21, 1989.

[25] J. Cao, J. Llibre, and X. Zhang. Darboux integrability and algebraic limit cycles for a class of polynomial differential systems. *Sci. China Math.*, 57(4):775–794, 2014.

[26] M. Caubergh and F. Dumortier. Hopf–Takens bifurcations and centres. *J. Differential Equations*, 202(1):1–31, 2004.

[27] M. Caubergh, F. Dumortier, and R. Roussarie. Alien limit cycles near a Hamiltonian 2-saddle cycle. *C. R. Math. Acad. Sci. Paris*, 340(8):587–592, 2005.

[28] M. Caubergh, F. Dumortier, and R. Roussarie. Alien limit cycles in rigid unfoldings of a Hamiltonian 2-saddle cycle. *Commun. Pure Appl. Anal.*, 6(1):1–21, 2007.

[29] X. Cen. New lower bound for the number of critical periods for planar polynomial systems. *J. Differential Equations*, 271:480–498, 2021.

[30] G. Chang, T. Zhang, and M. Han. On the number of limit cycles of a class of polynomial systems of Liénard type. *J. Math. Anal. Appl.*, 408(2):775–780, 2013.

[31] J. Chavarriga and M. Sabatini. A survey of isochronous centers. *Qual. Theory Dyn. Syst.*, 1(1):1–70, 1999.

[32] F. Chen, C. Li, J. Llibre, and Z. Zhang. A unified proof on the weak Hilbert 16th problem for $n = 2$. *J. Differential Equations*, 221(2):309–342, 2006.

[33] G. Chen, C. Li, C. Liu, and J. Llibre. The cyclicity of period annuli of some classes of reversible quadratic systems. *Discrete Contin. Dyn. Syst.*, 16(1):157–177, 2006.

[34] H. Chen and X. Chen. Dynamical analysis of a cubic Liénard system with global parameters. *Nonlinearity*, 28(10):3535–3562, 2015.

[35] H. Chen and X. Chen. Dynamical analysis of a cubic Liénard system with global parameters (II). *Nonlinearity*, 29(6):1798–1826, 2016.

[36] H. Chen and X. Chen. A proof of Wang–Kooij's conjectures for a cubic Liénard system with a cusp. *J. Math. Anal. Appl.*, 445(1):884–897, 2017.

[37] H. Chen and X. Chen. Global phase portraits of a degenerate Bogdanov–Takens system with symmetry (II). *Discrete Contin. Dyn. Syst. Ser. B*, 23(10):4141–4170, 2018.

[38] H. Chen, X. Chen, and J. Xie. Global phase portrait of a degenerate Bogdanov–Takens system with symmetry. *Discrete Contin. Dyn. Syst. Ser. B*, 22(4):1273–1293, 2017.

[39] L. Chen, X. Ma, G. Zhang, and C. Li. Cyclicity of several quadratic reversible systems with center of genus one. *J. Appl. Anal. Comput.*, 1(4):439–447, 2011.

[40] L. Chen and M. Wang. The relative position, and the number, of limit cycles of a quadratic differential system. *Acta Math. Sinica*, 22(6):751–758, 1979. (In Chinese).

[41] L. Chen and Y. Ye. Uniqueness of limit cycle of the systems of equations $dx/dt = -y + dx + lx^2 + xy + ny^2, dy/dt = x$. *Acta Math. Sinica*, 18:219–222, 1975. (In Chinese).

[42] X. Chen, W. Huang, V. G. Romanovski, and W. Zhang. Linearizability conditions of a time-reversible quartic-like system. *J. Math. Anal. Appl.*, 383(1):179–189, 2011.

[43] X. Chen, W. Huang, V. G. Romanovski, and W. Zhang. Linearizability and local bifurcation of critical periods in a cubic Kolmogorov system. *J. Comput. Appl. Math.*, 245:86–96, 2013.

[44] X. Chen, V. G. Romanovski, and W. Zhang. Critical periods of perturbations of reversible rigidly isochronous centers. *J. Differential Equations*, 251(6):1505–1525, 2011.

[45] X. Chen and W. Zhang. Weak centers of a cubic isochrone perturbed by fourth degree homogeneous polynomial. *Sichuan Daxue Xuebao*, 41(2):251–255, 2004.

[46] C. Chicone. The monotonicity of the period function for planar Hamiltonian vector fields. *J. Differential Equations*, 69(3):310–321, 1987.

[47] C. Chicone. *Ordinary differential equations with applications*, volume 34 of *Texts in Applied Mathematics*. Springer, New York, second edition, 2006.

[48] C. Chicone and F. Dumortier. A quadratic system with a nonmonotonic period function. *Proc. Amer. Math. Soc.*, 102(3):706–710, 1988.

[49] C. Chicone and F. Dumortier. Finiteness for critical periods of planar analytic vector fields. *Nonlinear Anal.*, 20(4):315–335, 1993.

[50] C. Chicone and M. Jacobs. Bifurcation of critical periods for plane vector fields. *Trans. Amer. Math. Soc.*, 312(2):433–486, 1989.

[51] C. Chicone and M. Jacobs. Bifurcation of limit cycles from quadratic isochrones. *J. Differential Equations*, 91(2):268–326, 1991.

[52] C. Chicone and J. Tian. On general properties of quadratic systems. *Amer. Math. Monthly*, 89(3):167–178, 1982.

[53] S.-N. Chow, C. Li, and D. Wang. *Normal forms and bifurcation of planar vector fields*. Cambridge University Press, Cambridge, 1994.

[54] S.-N. Chow, C. Li, and Y. Yi. The cyclicity of period annuli of degenerate quadratic Hamiltonian systems with elliptic segment loops. *Ergodic Theory Dynam. Systems*, 22(2):349–374, 2002.

[55] S.-N. Chow and J. A. Sanders. On the number of critical points of the period. *J. Differential Equations*, 64(1):51–66, 1986.

[56] C. Christopher. Estimating limit cycle bifurcations from centers. In *Differential equations with symbolic computation*, Trends Math., pages 23–35. Birkhäuser, Basel, 2005.

[57] C. Christopher and N. G. Lloyd. Polynomial systems: a lower bound for the Hilbert numbers. *Proc. Roy. Soc. London Ser. A*, 450(1938):219–224, 1995.

[58] C. Christopher and N. G. Lloyd. Small-amplitude limit cycles in polynomial Liénard systems. *NoDEA Nonlinear Differential Equations Appl.*, 3(2):183–190, 1996.

[59] A. Cima, A. Gasull, and F. Mañosas. Some applications of the Euler-Jacobi formula to differential equations. *Proc. Amer. Math. Soc.*, 118(1):151–163, 1993.

[60] A. Cima, A. Gasull, and F. Mañosas. Period function for a class of Hamiltonian systems. *J. Differential Equations*, 168(1):180–199, 2000. Special issue in celebration of Jack K. Hale's 70th birthday, Part 1 (Atlanta, GA/Lisbon, 1998).

[61] B. Coll, F. Dumortier, and R. Prohens. Alien limit cycles in Liénard equations. *J. Differential Equations*, 254(3):1582–1600, 2013.

[62] B. Coll, F. Dumortier, and R. Prohens. Configurations of limit cycles in Liénard equations. *J. Differential Equations*, 255(11):4169–4184, 2013.

[63] B. Coll, A. Gasull, and R. Prohens. First Lyapunov constants for non-smooth Liénard differential equations. In *Proceedings of the 2nd Catalan Days on Applied Mathematics (Odeillo, 1995)*, Collect. Études, pages 77–83. Presses Univ. Perpignan, Perpignan, 1995.

[64] B. Coll, A. Gasull, and R. Prohens. Bifurcation of limit cycles from two families of centers. *Dyn. Contin. Discrete Impuls. Syst. Ser. A Math. Anal.*, 12(2):275–287, 2005.

[65] B. Coll, C. Li, and R. Prohens. Quadratic perturbations of a class of quadratic reversible systems with two centers. *Discrete Contin. Dyn. Syst.*, 24(3):699–729, 2009.

[66] B. Coll and J. Llibre. Limit cycles for a quadratic system with an invariant straight line and some evolution of phase portraits. In *Qualitative theory of differential equations (Szeged, 1988)*, volume 53 of *Colloq. Math. Soc. János Bolyai*, pages 111–123. North-Holland, Amsterdam, 1990.

[67] W. A. Coppel. A survey of quadratic systems. *J. Differential Equations*, 2:293–304, 1966.

[68] W. A. Coppel and L. Gavrilov. The period function of a Hamiltonian quadratic system. *Differential Integral Equations*, 6(6):1357–1365, 1993.

[69] L. P. C. da Cruz, D. D. Novaes, and J. Torregrosa. New lower bound for the Hilbert number in piecewise quadratic differential systems. *J. Differential Equations*, 266(7):4170–4203, 2019.

[70] P. De Maesschalck and F. Dumortier. Classical Liénard equations of degree $n \geq 6$ can have $[\frac{n-1}{2}]+2$ limit cycles. *J. Differential Equations*, 250(4):2162–2176, 2011.

[71] P. De Maesschalck and R. Huzak. Slow divergence integrals in classical Liénard equations near centers. *J. Dynam. Differential Equations*, 27(1):177–185, 2015.

[72] P. De Maesschalck and J. Torregrosa. Critical periods in planar polynomial centers near a maximum number of cusps. Preprint, June 2023.

[73] T. Ding. *Applications of qualitative methods of ordinary differential equations*. Higher Education Press, Beijing, 2004. (In Chinese).

[74] H. Dulac. Sur les cycles limites. *Bull. Soc. Math. France*, 51:45–188, 1923.

[75] F. Dumortier, M. El Morsalani, and C. Rousseau. Hilbert's 16th problem for quadratic systems and cyclicity of elementary graphics. *Nonlinearity*, 9(5):1209–1261, 1996.

[76] F. Dumortier, A. Guzmán, and C. Rousseau. Finite cyclicity of elementary graphics surrounding a focus or center in quadratic systems. *Qual. Theory Dyn. Syst.*, 3(1):123–154, 2002.

[77] F. Dumortier, Y. Ilyashenko, and C. Rousseau. Normal forms near a saddle-node and applications to finite cyclicity of graphics. *Ergodic Theory Dynam. Systems*, 22(3):783–818, 2002.

[78] F. Dumortier and C. Li. Perturbations from an elliptic Hamiltonian of degree four. I. Saddle loop and two saddle cycle. *J. Differential Equations*, 176(1):114–157, 2001.

[79] F. Dumortier and C. Li. Perturbations from an elliptic Hamiltonian of degree four. II. Cuspidal loop. *J. Differential Equations*, 175(2):209–243, 2001.

[80] F. Dumortier and C. Li. Perturbation from an elliptic Hamiltonian of degree four. III. Global centre. *J. Differential Equations*, 188(2):473–511, 2003.

[81] F. Dumortier and C. Li. Perturbation from an elliptic Hamiltonian of degree four. IV. Figure eight-loop. *J. Differential Equations*, 188(2):512–554, 2003.

[82] F. Dumortier, C. Li, and Z. Zhang. Unfolding of a quadratic integrable system with two centers and two unbounded heteroclinic loops. *J. Differential Equations*, 139(1):146–193, 1997.

[83] F. Dumortier, D. Panazzolo, and R. Roussarie. More limit cycles than expected in Liénard equations. *Proc. Amer. Math. Soc.*, 135(6):1895–1904, 2007.

[84] F. Dumortier and R. Roussarie. Abelian integrals and limit cycles. *J. Differential Equations*, 227(1):116–165, 2006.

[85] F. Dumortier, R. Roussarie, and C. Rousseau. Elementary graphics of cyclicity 1 and 2. *Nonlinearity*, 7(3):1001–1043, 1994.

[86] F. Dumortier, R. Roussarie, and C. Rousseau. Hilbert's 16th problem for quadratic vector fields. *J. Differential Equations*, 110(1):86–133, 1994.

[87] F. Dumortier, R. Roussarie, and J. Sotomayor. Generic 3-parameter families of vector fields on the plane, unfolding a singularity with nilpotent linear part. The cusp case of codimension 3. *Ergodic Theory Dynam. Systems*, 7(3):375–413, 1987.

[88] J. Écalle. *Introduction aux fonctions analysables et preuve constructive de la conjecture de Dulac*. Actualités Mathématiques. [Current Mathematical Topics]. Hermann, Paris, 1992.

[89] W. W. Farr, C. Li, I. S. Labouriau, and W. F. Langford. Degenerate Hopf bifurcation formulas and Hilbert's 16th problem. *SIAM J. Math. Anal.*, 20(1):13–30, 1989.

[90] J.-P. Françoise. Successive derivatives of a first return map, application to the study of quadratic vector fields. *Ergodic Theory Dynam. Systems*, 16(1):87–96, 1996.

[91] J.-P. Françoise and C. C. Pugh. Keeping track of limit cycles. *J. Differential Equations*, 65(2):139–157, 1986.

[92] J.-P. Françoise and D. Xiao. Perturbation theory of a symmetric center within Liénard equations. *J. Differential Equations*, 259(6):2408–2429, 2015.

[93] E. Freire, A. Gasull, and A. Guillamon. Period function for perturbed isochronous centres. *Qual. Theory Dyn. Syst.*, 3(1):275–284, 2002.

[94] E. Freire, A. Gasull, and A. Guillamon. A characterization of isochronous centres in terms of symmetries. *Rev. Mat. Iberoamericana*, 20(1):205–222, 2004.

[95] A. Garijo and J. Villadelprat. Algebraic and analytical tools for the study of the period function. *J. Differential Equations*, 257(7):2464–2484, 2014.

[96] A. Gasull, H. Giacomini, S. Pérez-González, and J. Torregrosa. A proof of Perko's conjectures for the Bogdanov-Takens system. *J. Differential Equations*, 255(9):2655–2671, 2013.

[97] A. Gasull, H. Giacomini, and J. Torregrosa. Some results on homoclinic and heteroclinic connections in planar systems. *Nonlinearity*, 23(12):2977–3001, 2010.

[98] A. Gasull and A. Guillamon. Non-existence, uniqueness of limit cycles and center problem in a system that includes predator-prey systems and generalized Liénard equations. *Differential Equations Dynam. Systems*, 3(4):345–366, 1995.

[99] A. Gasull, A. Guillamon, and J. Villadelprat. The period function for second-order quadratic ODEs is monotone. *Qual. Theory Dyn. Syst.*, 4(2):329–352 (2004), 2003.

[100] A. Gasull, J. T. Lázaro, and J. Torregrosa. Upper bounds for the number of zeroes for some Abelian integrals. *Nonlinear Anal.*, 75(13):5169–5179, 2012.

[101] A. Gasull, C. Li, and J. Torregrosa. A new Chebyshev family with applications to Abel equations. *J. Differential Equations*, 252(2):1635–1641, 2012.

[102] A. Gasull, C. Li, and J. Torregrosa. Limit cycles for 3-monomial differential equations. *J. Math. Anal. Appl.*, 428(2):735–749, 2015.

[103] A. Gasull, W. Li, J. Llibre, and Z. Zhang. Chebyshev property of complete elliptic integrals and its application to Abelian integrals. *Pacific J. Math.*, 202(2):341–361, 2002.

[104] A. Gasull, C. Liu, and J. Yang. On the number of critical periods for planar polynomial systems of arbitrary degree. *J. Differential Equations*, 249(3):684–692, 2010.

[105] A. Gasull, V. Mañosa, and J. Villadelprat. On the period of the limit cycles appearing in one-parameter bifurcations. *J. Differential Equations*, 213(2):255–288, 2005.

[106] A. Gasull and J. Torregrosa. Small-amplitude limit cycles in Liénard systems via multiplicity. *J. Differential Equations*, 159(1):186–211, 1999.

[107] A. Gasull and J. Torregrosa. A new algorithm for the computation of the Lyapunov constants for some degenerated critical points. *Nonlinear Anal.*, 47(7):4479–4490, 2001.

[108] A. Gasull and J. Torregrosa. A new approach to the computation of the Lyapunov constants. *Comput. Appl. Math.*, 20(1-2):149–177, 2001.

[109] A. Gasull and J. Torregrosa. A relation between small amplitude and big limit cycles. *Rocky Mountain J. Math.*, 31(4):1277–1303, 2001.

[110] S. Gautier, L. Gavrilov, and I. D. Iliev. Perturbations of quadratic centers of genus one. *Discrete Contin. Dyn. Syst.*, 25(2):511–535, 2009.

[111] L. Gavrilov. Remark on the number of critical points of the period. *J. Differential Equations*, 101(1):58–65, 1993.

[112] L. Gavrilov. Petrov modules and zeros of Abelian integrals. *Bull. Sci. Math.*, 122(8):571–584, 1998.

[113] L. Gavrilov. The infinitesimal 16th Hilbert problem in the quadratic case. *Invent. Math.*, 143(3):449–497, 2001.

[114] L. Gavrilov and I. D. Iliev. Second-order analysis in polynomially perturbed reversible quadratic Hamiltonian systems. *Ergodic Theory Dynam. Systems*, 20(6):1671–1686, 2000.

[115] L. Gavrilov and I. D. Iliev. Bifurcations of limit cycles from infinity in quadratic systems. *Canad. J. Math.*, 54(5):1038–1064, 2002.

[116] L. Gavrilov and I. D. Iliev. Two-dimensional Fuchsian systems and the Chebyshev property. *J. Differential Equations*, 191(1):105–120, 2003.

[117] L. Gavrilov and I. D. Iliev. The displacement map associated to polynomial unfoldings of planar Hamiltonian vector fields. *Amer. J. Math.*, 127(6):1153–1190, 2005.

[118] L. Gavrilov and I. D. Iliev. Cubic perturbations of elliptic Hamiltonian vector fields of degree three. *J. Differential Equations*, 260(5):3963–3990, 2016.

[119] J. Giné. Higher order limit cycle bifurcations from non-degenerate centers. *Appl. Math. Comput.*, 218(17):8853–8860, 2012.

[120] J. Giné. Limit cycle bifurcations from a non-degenerate center. *Appl. Math. Comput.*, 218(9):4703–4709, 2012.

[121] J. Giné, L. F. S. Gouveia, and J. Torregrosa. Lower bounds for the local cyclicity for families of centers. *J. Differential Equations*, 275:309–331, 2021.

[122] J. Giné, M. Grau, and J. Llibre. Averaging theory at any order for computing periodic orbits. *Phys. D*, 250:58–65, 2013.

[123] J. Giné, M. Grau, and J. Llibre. Limit cycles bifurcating from planar polynomial quasi-homogeneous centers. *J. Differential Equations*, 259(12):7135–7160, 2015.

[124] J. Giné and J. Llibre. Limit cycles of cubic polynomial vector fields via the averaging theory. *Nonlinear Anal.*, 66(8):1707–1721, 2007.

[125] J. Giné, J. Llibre, K. Wu, and X. Zhang. Averaging methods of arbitrary order, periodic solutions and integrability. *J. Differential Equations*, 260(5):4130–4156, 2016.

[126] J. Giné and X. Santallusia. Implementation of a new algorithm of computation of the Poincaré-Liapunov constants. *J. Comput. Appl. Math.*, 166(2):465–476, 2004.

[127] L. F. S. Gouveia and J. Torregrosa. Lower bounds for the local cyclicity of centers using high order developments and parallelization. *J. Differential Equations*, 271:447–479, 2021.

[128] L. F. S. Gouveia and J. Torregrosa. The local cyclicity problem: Melnikov method using Lyapunov constants. *Proc. Edinb. Math. Soc. (2)*, 65(2):356–375, 2022.

[129] M. Grau, F. Mañosas, and J. Villadelprat. A Chebyshev criterion for Abelian integrals. *Trans. Amer. Math. Soc.*, 363(1):109–129, 2011.

[130] M. Han. Bifurcation theory of limit cycles of planar systems. In *Handbook of differential equations: ordinary differential equations. Vol. III*, Handb. Differ. Equ., pages 341–433. Elsevier/North-Holland, Amsterdam, 2006.

[131] M. Han. Asymptotic expansions of Melnikov functions and limit cycle bifurcations. *Internat. J. Bifur. Chaos Appl. Sci. Engrg.*, 22(12):1250296, 30, 2012.

[132] M. Han. *Bifurcation theory of limit cycles.* Science Press Beijing, Beijing; Alpha Science International Ltd., Oxford, 2017.

[133] M. Han, J. Jiang, and H. Zhu. Limit cycle bifurcations in near-Hamiltonian systems by perturbing a nilpotent center. *Internat. J. Bifur. Chaos Appl. Sci. Engrg.*, 18(10):3013–3027, 2008.

[134] M. Han and J. Li. Lower bounds for the Hilbert number of polynomial systems. *J. Differential Equations*, 252(4):3278–3304, 2012.

[135] M. Han, J. Llibre, and J. Yang. On uniqueness of limit cycles in general Bogdanov–Takens bifurcation. *Internat. J. Bifur. Chaos Appl. Sci. Engrg.*, 28(9):1850115, 12, 2018.

[136] M. Han, V. G. Romanovski, and X. Zhang. Equivalence of the Melnikov function method and the averaging method. *Qual. Theory Dyn. Syst.*, 15(2):471–479, 2016.

[137] M. Han, L. Sheng, and X. Zhang. Bifurcation theory for finitely smooth planar autonomous differential systems. *J. Differential Equations*, 264(5):3596–3618, 2018.

[138] M. Han and Y. Xiong. Limit cycle bifurcations in a class of near-Hamiltonian systems with multiple parameters. *Chaos Solitons Fractals*, 68:20–29, 2014.

[139] M. Han, J. Yang, A.-A. Tarţa, and Y. Gao. Limit cycles near homoclinic and heteroclinic loops. *J. Dynam. Differential Equations*, 20(4):923–944, 2008.

[140] M. Han, J. Yang, and D. Xiao. Limit cycle bifurcations near a double homoclinic loop with a nilpotent saddle. *Internat. J. Bifur. Chaos Appl. Sci. Engrg.*, 22(8):1250189, 33, 2012.

[141] M. Han and P. Yu. *Normal forms, Melnikov functions and bifurcations of limit cycles*, volume 181 of *Applied Mathematical Sciences*. Springer, London, 2012.

[142] M. Han, T. Zhang, and H. Zang. On the number and distribution of limit cycles in a cubic system. *Internat. J. Bifur. Chaos Appl. Sci. Engrg.*, 14(12):4285–4292, 2004.

[143] M. Han, T. Zhang, and H. Zang. Bifurcation of limit cycles near equivariant compound cycles. *Sci. China Ser. A*, 50(4):503–514, 2007.

[144] M. Han and Z. Zhang. Cyclicity 1 and 2 conditions for a 2-polycycle of integrable systems on the plane. *J. Differential Equations*, 155(2):245–261, 1999.

[145] M. Han and D. Zhu. *Bifurcation theory of differential equations*. Coal Mine Industry Publishing House, 1994. (In Chinese).

[146] M. Han and H. Zhu. The loop quantities and bifurcations of homoclinic loops. *J. Differential Equations*, 234(2):339–359, 2007.

[147] D. Hilbert. Mathematical problems. *Bull. Amer. Math. Soc. (N.S.)*, 37(4):407–436, 2000. Reprinted from Bull. Amer. Math. Soc. **8** (1902), 437–479.

[148] X. Hong, J. Lu, and Y. Wang. Upper bounds for the associated number of zeros of Abelian integrals for two classes of quadratic reversible centers of genus one. *J. Appl. Anal. Comput.*, 8(6):1959–1970, 2018.

[149] X. Hong, S. Xie, and L. Chen. Estimating the number of zeros for Abelian integrals of quadratic reversible centers with orbits formed by higher-order curves. *Internat. J. Bifur. Chaos Appl. Sci. Engrg.*, 26(2):1650020, 16, 2016.

[150] X. Hong, S. Xie, and R. Ma. On the Abelian integrals of quadratic reversible centers with orbits formed by genus one curves of higher degree. *J. Math. Anal. Appl.*, 429(2):924–941, 2015.

[151] E. Horozov. Versal deformations of equivariant vector fields for cases of symmetry of order 2 and 3. *Trudy Sem. Petrovsk.*, 5:163–192, 1979.

[152] E. Horozov and I. D. Iliev. Hilbert–Arnold problem for cubic Hamiltonians and limit cycles. In *Proceedings of the Fourth International Colloquium on Differential Equations (Plovdiv, 1993)*, pages 115–124. VSP, Utrecht, 1994.

[153] E. Horozov and I. D. Iliev. On saddle-loop bifurcations of limit cycles in perturbations of quadratic Hamiltonian systems. *J. Differential Equations*, 113(1):84–105, 1994.

[154] E. Horozov and I. D. Iliev. On the number of limit cycles in perturbations of quadratic Hamiltonian systems. *Proc. London Math. Soc. (3)*, 69(1):198–224, 1994.

[155] R. Huzak. Cyclicity of degenerate graphic DF_{2a} of Dumortier-Roussarie-Rousseau program. *Commun. Pure Appl. Anal.*, 17(3):1305–1316, 2018.

[156] I. D. Iliev. The cyclicity of the period annulus of the quadratic Hamiltonian triangle. *J. Differential Equations*, 128(1):309–326, 1996.

[157] I. D. Iliev. Higher-order Melnikov functions for degenerate cubic Hamiltonians. *Adv. Differential Equations*, 1(4):689–708, 1996.

[158] I. D. Iliev. On second order bifurcations of limit cycles. *J. London Math. Soc. (2)*, 58(2):353–366, 1998.

[159] I. D. Iliev. Perturbations of quadratic centers. *Bull. Sci. Math.*, 122(2):107–161, 1998.

[160] I. D. Iliev. On the limit cycles available from polynomial perturbations of the Bogdanov–Takens Hamiltonian. *Israel J. Math.*, 115:269–284, 2000.

[161] I. D. Iliev, C. Li, and J. Yu. Bifurcations of limit cycles from quadratic non-Hamiltonian systems with two centres and two unbounded heteroclinic loops. *Nonlinearity*, 18(1):305–330, 2005.

[162] I. D. Iliev, C. Li, and J. Yu. On the cubic perturbations of the symmetric 8-loop Hamiltonian. *J. Differential Equations*, 269(4):3387–3413, 2020.

[163] I. D. Iliev and L. M. Perko. Higher order bifurcations of limit cycles. *J. Differential Equations*, 154(2):339–363, 1999.

[164] Y. Ilyashenko. The appearance of limit cycles under a perturbation of the equation $dw/dz = -R_z/R_w$, where $R(z, w)$ is a polynomial. *Mat. Sb. (N.S.)*, 78 (120):360–373, 1969.

[165] Y. Ilyashenko. The multiplicity of limit cycles that arise in the perturbation of a Hamiltonian equation of the class $\omega' = P_2/Q_1$ in a real and complex domain. *Trudy Sem. Petrovsk.*, 3:49–60, 1978.

[166] Y. Ilyashenko. Finiteness theorems for limit cycles. *Uspekhi Mat. Nauk*, 45(2(272)):143–200, 240, 1990.

[167] Y. Ilyashenko. Centennial history of Hilbert's 16th problem. *Bull. Amer. Math. Soc. (N.S.)*, 39(3):301–354, 2002.

[168] Y. Ilyashenko. Notes of the advanced course "Recent Trends in Nonlinear Science". Centre de Recerca Matemàtica, Barcelona, February 3–6, 2020.

[169] Y. Ilyashenko and W. Li. *Nonlocal bifurcations*, volume 66 of *Mathematical Surveys and Monographs*. American Mathematical Society, Providence, RI, 1999.

[170] Y. Ilyashenko and S. Yakovenko. Double exponential estimate for the number of zeros of complete Abelian integrals and rational envelopes of linear ordinary differential equations with an irreducible monodromy group. *Invent. Math.*, 121(3):613–650, 1995.

[171] Y. Ilyashenko and S. Yakovenko. *Lectures on analytic differential equations*, volume 86 of *Graduate Studies in Mathematics*. American Mathematical Society, Providence, RI, 2008.

[172] X. Jarque and J. Villadelprat. Nonexistence of isochronous centers in planar polynomial Hamiltonian systems of degree four. *J. Differential Equations*, 180(2):334–373, 2002.

[173] S. Karlin and W. J. Studden. *Tchebycheff systems: With applications in analysis and statistics*. Pure and Applied Mathematics, Vol. XV. Interscience Publishers John Wiley & Sons, New York-London-Sydney, 1966.

[174] A. G. Khovanskiĭ. Real analytic manifolds with the property of finiteness, and complex Abelian integrals. *Funktsional. Anal. i Prilozhen.*, 18(2):40–50, 1984.

[175] R. E. Kooij and A. Zegeling. Limit cycles in quadratic systems with a weak focus and a strong focus. *Kyungpook Math. J.*, 38(2):323–340, 1998.

[176] B. Li and Z. Zhang. A note on a result of G. S. Petrov about the weakened 16th Hilbert problem. *J. Math. Anal. Appl.*, 190(2):489–516, 1995.

[177] C. Li. Two problems of planar quadratic systems. *Sci. Sinica Ser. A*, 26(5):471–481, 1983.

[178] C. Li. Nonexistence of limit cycle around a weak focus of order three for any quadratic system. *Chinese Ann. Math. Ser. B*, 7(2):174–190, 1986. A Chinese summary appears in Chinese Ann. Math. Ser. A **7** (1986), no. 2, 239.

[179] C. Li. On the proof of a theorem about the monotonicity of ratio of two Abelian integrals, 1996. Appendix of the paper "Canard Cycles and Center Manifols" by F. Dumortier and R. Roussarie, Mem. Amer. Math. Soc. 121, No. 577 (1996).

[180] C. Li. Abelian integrals and limit cycles. *Qual. Theory Dyn. Syst.*, 11(1):111–128, 2012.

[181] C. Li. Study of period functions. *Math. Theory Appl.*, 43(1):1–31, 2023.

[182] C. Li, J. Li, and Z. Ma. Codimension 3 B-T bifurcations in an epidemic model with a nonlinear incidence. *Discrete Contin. Dyn. Syst. Ser. B*, 20(4):1107–1116, 2015.

[183] C. Li and W. Li. Weak Hilbert's 16th problem and the related research. *Adv. Math. (China)*, 39(5):513–526, 2010.

[184] C. Li and W. Li. The uniqueness of limit cycles for classical Liénard equations of degree four under singular perturbations. *Sci. Sin. Math.*, 47(1):119–134, 2017. (In Chinese).

[185] C. Li, W. Li, J. Llibre, and Z. Zhang. Linear estimate for the number of zeros of Abelian integrals for quadratic isochronous centres. *Nonlinearity*, 13(5):1775–1800, 2000.

[186] C. Li, W. Li, J. Llibre, and Z. Zhang. Linear estimation of the number of zeros of Abelian integrals for some cubic isochronous centers. *J. Differential Equations*, 180(2):307–333, 2002.

[187] C. Li, C. Liu, and J. Yang. A cubic system with thirteen limit cycles. *J. Differential Equations*, 246(9):3609–3619, 2009.

[188] C. Li and J. Llibre. A unified study on the cyclicity of period annulus of the reversible quadratic Hamiltonian systems. *J. Dynam. Differential Equations*, 16(2):271–295, 2004.

[189] C. Li and J. Llibre. Quadratic perturbations of a quadratic reversible Lotka–Volterra system. *Qual. Theory Dyn. Syst.*, 9(1-2):235–249, 2010.

[190] C. Li and J. Llibre. Uniqueness of limit cycles for Liénard differential equations of degree four. *J. Differential Equations*, 252(4):3142–3162, 2012.

[191] C. Li, J. Llibre, and Z. Zhang. Weak focus, limit cycles, and bifurcations for bounded quadratic systems. *J. Differential Equations*, 115(1):193–223, 1995.

[192] C. Li and K. Lu. The period function of hyperelliptic Hamiltonians of degree 5 with real critical points. *Nonlinearity*, 21(3):465–483, 2008. Supplementary data files are available from the article's abstract page in the online journal.

[193] C. Li, Z. Ma, and Y. Zhou. Periodic orbits in 3-dimensional systems and application to a perturbed Volterra system. *J. Differential Equations*, 260(3):2750–2762, 2016.

[194] C. Li, P. Mardešić, and R. Roussarie. Perturbations of symmetric elliptic Hamiltonians of degree four. *J. Differential Equations*, 231(1):78–91, 2006.

[195] C. Li and R. Roussarie. The cyclicity of the elliptic segment loops of the reversible quadratic Hamiltonian systems under quadratic perturbations. *J. Differential Equations*, 205(2):488–520, 2004.

[196] C. Li and C. Rousseau. A system with three limit cycles appearing in a Hopf bifurcation and dying in a homoclinic bifurcation: the cusp of order 4. *J. Differential Equations*, 79(1):132–167, 1989.

[197] C. Li, C. Rousseau, and X. Wang. A simple proof for the unicity of the limit cycle in the Bogdanov-Takens system. *Canad. Math. Bull.*, 33(1):84–92, 1990.

[198] C. Li and Z. Zhang. A criterion for determining the monotonicity of the ratio of two Abelian integrals. *J. Differential Equations*, 124(2):407–424, 1996.

[199] C. Li and Z. Zhang. Remarks on 16th weak Hilbert problem for $n = 2$. *Nonlinearity*, 15(6):1975–1992, 2002.

[200] F. Li, Y. Liu, and Y. Jin. Bifurcations of limit circles and center conditions for a class of non-analytic cubic Z_2 polynomial differential systems. *Acta Math. Sin. (Engl. Ser.)*, 28(11):2275–2288, 2012.

[201] F. Li, Y. Liu, Y. Liu, and P. Yu. Bi-center problem and bifurcation of limit cycles from nilpotent singular points in Z_2-equivariant cubic vector fields. *J. Differential Equations*, 265(10):4965–4992, 2018.

[202] F. Li, Y. Liu, and Y. Wu. Center conditions and bifurcation of limit cycles at three-order nilpotent critical point in a seventh degree Lyapunov system. *Commun. Nonlinear Sci. Numer. Simul.*, 16(6):2598–2608, 2011.

[203] J. Li. Hilbert's 16th problem and bifurcations of planar polynomial vector fields. *Internat. J. Bifur. Chaos Appl. Sci. Engrg.*, 13(1):47–106, 2003.

[204] J. Li, H. S. Y. Chan, and K. W. Chung. Bifurcations of limit cycles in a Z_6-equivariant planar vector field of degree 5. *Sci. China Ser. A*, 45(7):817–826, 2002.

[205] J. Li and Q. Huang. Bifurcations of limit cycles forming compound eyes in the cubic system. *Chinese Ann. Math. Ser. B*, 8(4):391–403, 1987. A Chinese summary appears in Chinese Ann. Math. Ser. A **8** (1987), no. 5, 643.

[206] J. Li, C. Li, C. Liu, and D. Wang. The period function of reversible Lotka-Volterra quadratic centers. *J. Differential Equations*, 307:556–579, 2022.

[207] J. Li and C. F. Li. Planar cubic Hamiltonian systems and distributions of limit cycles of (E_3). *Acta Math. Sinica*, 28(4):509–521, 1985.

[208] J. Li and Y. Liu. New results on the study of Z_q-equivariant planar polynomial vector fields. *Qual. Theory Dyn. Syst.*, 9(1-2):167–219, 2010.

[209] J. Li and Z. Liu. Bifurcation set and limit cycles forming compound eyes in a perturbed Hamiltonian system. *Publ. Mat.*, 35(2):487–506, 1991.

[210] J. Li and S. Wan. Global bifurcations in a disturbed Hamiltonian vector field approaching a $3 : 1$ resonant Poincaré map. I. *Acta Math. Appl. Sinica (English Ser.)*, 7(1):80–89, 1991.

[211] J. Li, M. Zhang, and S. Li. Bifurcations of limit cycles in a Z_2-equivariant planar polynomial vector field of degree 7. *Internat. J. Bifur. Chaos Appl. Sci. Engrg.*, 16(4):925–943, 2006.

[212] J. Li and X. Zhao. Rotation symmetry groups of planar Hamiltonian systems. *Ann. Differential Equations*, 5(1):25–33, 1989.

[213] J. Li and H. Zhou. On the control of parameters of distributions of limit cycles for a Z_2-equivariant perturbed planar Hamiltonian polynomial vector field. *Internat. J. Bifur. Chaos Appl. Sci. Engrg.*, 15(1):137–155, 2005.

[214] W. Li. *Theory of normal forms and its applications.* Science Press Beijing, Beijing, 2000. (In Chinese).

[215] W. Li, Y. Zhao, C. Li, and Z. Zhang. Abelian integrals for quadratic centres having almost all their orbits formed by quartics. *Nonlinearity*, 15(3):863–885, 2002.

[216] H. Liang and J. Torregrosa. Parallelization of the Lyapunov constants and cyclicity for centers of planar polynomial vector fields. *J. Differential Equations*, 259(11):6494–6509, 2015.

[217] H. Liang and Y. Zhao. Quadratic perturbations of a class of quadratic reversible systems with one center. *Discrete Contin. Dyn. Syst.*, 27(1):325–335, 2010.

[218] A. Lins, W. de Melo, and C. C. Pugh. On Liénard's equation. In *Geometry and topology (Proc. III Latin Amer. School of Math., Inst. Mat. Pura Aplicada CNPq, Rio de Janeiro, 1976)*, pages 335–357. Lecture Notes in Math., Vol. 597, 1977.

[219] C. Liu. Estimate of the number of zeros of Abelian integrals for an elliptic Hamiltonian with figure-of-eight loop. *Nonlinearity*, 16(3):1151–1163, 2003.

[220] C. Liu. The cyclicity of period annuli of a class of quadratic reversible systems with two centers. *J. Differential Equations*, 252(10):5260–5273, 2012.

[221] C. Liu. Limit cycles bifurcated from some reversible quadratic centres with a non-algebraic first integral. *Nonlinearity*, 25(6):1653–1660, 2012.

[222] C. Liu, G. Chen, and Z. Sun. New criteria for the monotonicity of the ratio of two Abelian integrals. *J. Math. Anal. Appl.*, 465(1):220–234, 2018.

[223] C. Liu and D. Xiao. The monotonicity of the ratio of two Abelian integrals. *Trans. Amer. Math. Soc.*, 365(10):5525–5544, 2013.

[224] C. Liu and D. Xiao. The smallest upper bound on the number of zeros of Abelian integrals. *J. Differential Equations*, 269(4):3816–3852, 2020.

[225] Y. Liu. The values of singular point of E_n and some kinds of problems of bifurcation. *Sci. China Ser. A*, 36(5):550–560, 1993.

[226] Y. Liu. Theory of center-focus for a class of higher-degree critical points and infinite points. *Sci. China Ser. A*, 44(3):365–377, 2001.

[227] Y. Liu and H. Chen. Formulas of singular point quantities and the first 10 saddle quantities for a class of cubic system. *Acta Math. Appl. Sin.*, 25(2):295–302, 2002.

[228] Y. Liu and W. Huang. A cubic system with twelve small amplitude limit cycles. *Bull. Sci. Math.*, 129(2):83–98, 2005.

[229] Y. Liu and J. Li. Theory of values of singular point in complex autonomous differential systems. *Sci. China Ser. A*, 33(1):10–23, 1990.

[230] Y. Liu and J. Li. Center problem and multiple Hopf bifurcation for the Z_5-equivariant planar polynomial vector fields of degree 5. *Internat. J. Bifur. Chaos Appl. Sci. Engrg.*, 19(6):2115–2121, 2009.

[231] Y. Liu and J. Li. Center problem and multiple Hopf bifurcation for the Z_6-equivariant planar polynomial vector fields of degree 5. *Internat. J. Bifur. Chaos Appl. Sci. Engrg.*, 19(5):1741–1749, 2009.

[232] Y. Liu and J. Li. New study on the center problem and bifurcations of limit cycles for the Lyapunov system. I. *Internat. J. Bifur. Chaos Appl. Sci. Engrg.*, 19(11):3791–3801, 2009.

[233] Y. Liu and J. Li. New study on the center problem and bifurcations of limit cycles for the Lyapunov system. II. *Internat. J. Bifur. Chaos Appl. Sci. Engrg.*, 19(9):3087–3099, 2009.

[234] Y. Liu and J. Li. Z_2-equivariant cubic system which yields 13 limit cycles. *Acta Math. Appl. Sin. Engl. Ser.*, 30(3):781–800, 2014.

[235] Y. Liu, J. Li, and W. Huang. *Singular points values, center problem and bifurcations of two dimensional differential autonomous systems.* Science Press Beijing, Beijing, 2008. (In Chinese).

[236] J. Llibre. Averaging theory and limit cycles for quadratic systems. *Rad. Mat.*, 11(2):215–228, 2002/03.

[237] J. Llibre. Integrability of polynomial differential systems. In *Handbook of differential equations*, pages 437–532. Elsevier/North-Holland, Amsterdam, 2004.

[238] J. Llibre, A. C. Mereu, and D. D. Novaes. Averaging theory for discontinuous piecewise differential systems. *J. Differential Equations*, 258(11):4007–4032, 2015.

[239] J. Llibre, J. S. Pérez del Río, and J. A. Rodríguez. Averaging analysis of a perturbated quadratic center. *Nonlinear Anal.*, 46(1, Ser. A: Theory Methods):45–51, 2001.

[240] J. Llibre, L. Pizarro, and E. Ponce. Limit cycles of polynomial Liénard systems. Comment on: "Number of limit cycles of the Liénard equation" [Phys. Rev. E (3) **56** (1997), no. 4, 3809–3813; MR1476640 (98f:34035)] by H. Giacomini and S. Neukirch. *Phys. Rev. E (3)*, 58(4):5185–5187, 1998.

[241] J. Llibre and G. Rodríguez. Configurations of limit cycles and planar polynomial vector fields. *J. Differential Equations*, 198(2):374–380, 2004.

[242] J. Llibre and D. Schlomiuk. The geometry of quadratic differential systems with a weak focus of third order. *Canad. J. Math.*, 56(2):310–343, 2004.

[243] J. Llibre and X. Zhang. On the algebraic limit cycles of Liénard systems. *Nonlinearity*, 21(9):2011–2022, 2008.

[244] J. Llibre and X. Zhang. On the Hopf-zero bifurcation of the Michelson system. *Nonlinear Anal. Real World Appl.*, 12(3):1650–1653, 2011.

[245] J. Llibre and X. Zhang. Limit cycles of the classical Liénard differential systems: a survey on the Lins Neto, de Melo and Pugh's conjecture. *Expo. Math.*, 35(3):286–299, 2017.

[246] J. Llibre and X. Zhang. The non-existence, existence and uniqueness of limit cycles for quadratic polynomial differential systems. *Proc. Roy. Soc. Edinburgh Sect. A*, 149(1):1–14, 2019.

[247] N. G. Lloyd. Limit cycles of polynomial systems, some recent developments. In *New directions in dynamical systems*, volume 127 of *London Math. Soc. Lecture Note Ser.*, pages 192–234. Cambridge Univ. Press, Cambridge, 1988.

[248] N. G. Lloyd and S. Lynch. Small-amplitude limit cycles of certain Liénard systems. *Proc. Roy. Soc. London Ser. A*, 418(1854):199–208, 1988.

[249] W. S. Loud. Behavior of the period of solutions of certain plane autonomous systems near centers. *Contributions to Differential Equations*, 3:21–36, 1964.

[250] S. Luca, F. Dumortier, M. Caubergh, and R. Roussarie. Detecting alien limit cycles near a Hamiltonian 2-saddle cycle. *Discrete Contin. Dyn. Syst.*, 25(4):1081–1108, 2009.

[251] D. Luo, X. Wang, D. Zhu, and M. Han. *Bifurcation theory and methods of dynamical systems*, volume 15 of *Advanced Series in Dynamical Systems*. World Scientific Publishing Co., Inc., River Edge, NJ, 1997.

[252] F. Mañosas and J. Villadelprat. Bounding the number of zeros of certain Abelian integrals. *J. Differential Equations*, 251(6):1656–1669, 2011.

[253] P. Mardešić. The number of limit cycles of polynomial deformations of a Hamiltonian vector field. *Ergodic Theory Dynam. Systems*, 10(3):523–529, 1990.

[254] P. Mardešić. An explicit bound for the multiplicity of zeros of generic Abelian integrals. *Nonlinearity*, 4(3):845–852, 1991.

[255] P. Mardešić, D. Marín, and J. Villadelprat. The period function of reversible quadratic centers. *J. Differential Equations*, 224(1):120–171, 2006.

[256] P. Mardešić. *Chebyshev systems and the versal unfolding of the cusps of order n*, volume 57 of *Travaux en Cours [Works in Progress]*. Hermann, Paris, 1998.

[257] D. Marín and J. Villadelprat. On the Chebyshev property of certain Abelian integrals near a polycycle. *Qual. Theory Dyn. Syst.*, 17(1):261–270, 2018.

[258] Y. Markov. Limit cycles of perturbations of a class of quadratic Hamiltonian vector fields. *Serdica Math. J.*, 22(2):91–108, 1996.

[259] D. D. Novaes and J. Torregrosa. On extended Chebyshev systems with positive accuracy. *J. Math. Anal. Appl.*, 448(1):171–186, 2017.

[260] D. Novikov and S. Yakovenko. Simple exponential estimate for the number of real zeros of complete Abelian integrals. *Ann. Inst. Fourier (Grenoble)*, 45(4):897–927, 1995.

[261] D. Novikov and S. Yakovenko. Tangential Hilbert problem for perturbations of hyperelliptic Hamiltonian systems. *Electron. Res. Announc. Amer. Math. Soc.*, 5:55–65, 1999.

[262] N. F. Otrokov. On the number of limit cycles of a differential equation in the neighborhood of a singular point. *Mat. Sbornik N.S.*, 34(76):127–144, 1954.

[263] L. Peng. Unfolding of a quadratic integrable system with a homoclinic loop. *Acta Math. Sin. (Engl. Ser.)*, 18(4):737–754, 2002.

[264] L. Peng. Quadratic perturbations of a quadratic reversible center of genus one. *Front. Math. China*, 6(5):911–930, 2011.

[265] L. Peng and Z. Feng. Bifurcation of limit cycles from a quintic center via the second order averaging method. *Internat. J. Bifur. Chaos Appl. Sci. Engrg.*, 25(3):1550047, 18, 2015.

[266] L. Peng and Z. Feng. Limit cycles from a cubic reversible system via the third-order averaging method. *Electron. J. Differential Equations*, pages No. 111, 27, 2015.

[267] L. Peng, Z. Feng, and C. Liu. Quadratic perturbations of a quadratic reversible Lotka–Volterra system with two centers. *Discrete Contin. Dyn. Syst.*, 34(11):4807–4826, 2014.

[268] L. Peng and Y. Lei. The cyclicity of the period annulus of a quadratic reversible system with a hemicycle. *Discrete Contin. Dyn. Syst.*, 30(3):873–890, 2011.

[269] L. Peng and Y. Lei. Bifurcation of limit cycles from a quadratic reversible center with the unbounded elliptic separatrix. *Bull. Iranian Math. Soc.*, 39(6):1223–1248, 2013.

[270] L. Peng and Y. Li. On the limit cycles bifurcating from a quadratic reversible center of genus one. *Mediterr. J. Math.*, 11(2):373–392, 2014.

[271] L. Peng and Y. Sun. The cyclicity of the period annulus of a quadratic reversible system with one center of genus one. *Turkish J. Math.*, 35(4):667–685, 2011.

[272] D. Peralta-Salas. Note on a paper of J. Llibre and G. Rodríguez concerning algebraic limit cycles: "Configurations of limit cycles and planar polynomial vector fields" [J. Differential Equations **198** (2004), no. 2, 374–380; MR2039147]. *J. Differential Equations*, 217(1):249–256, 2005.

[273] L. M. Perko. A global analysis of the Bogdanov-Takens system. *SIAM J. Appl. Math.*, 52(4):1172–1192, 1992.

[274] G. S. Petrov. The number of zeros of complete elliptic integrals. *Funktsional. Anal. i Prilozhen.*, 18(2):73–74, 1984.

[275] G. S. Petrov. Elliptic integrals and their nonoscillation. *Funktsional. Anal. i Prilozhen.*, 20(1):46–49, 96, 1986.

[276] G. S. Petrov. The Chebyshev property of elliptic integrals. *Funktsional. Anal. i Prilozhen.*, 22(1):83–84, 1988.

[277] G. S. Petrov. Nonoscillation of elliptic integrals. *Funktsional. Anal. i Prilozhen.*, 24(3):45–50, 96, 1990.

[278] G. S. Petrov. On the nonoscillation of elliptic integrals. *Funktsional. Anal. i Prilozhen.*, 31(4):47–51, 95, 1997.

[279] I. G. Petrovskiĭ and E. M. Landis. On the number of limit cycles of the equation $dy/dx = P(x, y)/Q(x, y)$, where P and Q are polynomials of the second degree. In *American Mathematical Society Translations, Ser. 2, Vol. 10*, pages 177–221. American Mathematical Society, Providence, R.I., 1958.

[280] H. Poincaré. Sur le problème des trois corps et les équations de la dynamique. *Acta Math.*, 13(1-2):1–270, 1890.

[281] L. Pontryagin. On dynamical systems close to Hamiltonian ones. *Zh. Exp. & Theor. Phys.*, 4:234–238, 1934.

[282] R. Prohens and J. Torregrosa. Corrigendum to "Shape and period of limit cycles bifurcating from a class of Hamiltonian period annulus" [Nonlinear Anal. 81 (2013) 130–148]. *Nonlinear Anal.*, 93:1–2, 2013.

[283] R. Prohens and J. Torregrosa. Shape and period of limit cycles bifurcating from a class of Hamiltonian period annulus. *Nonlinear Anal.*, 81:130–148, 2013.

[284] R. Prohens and J. Torregrosa. Periodic orbits from second order perturbation via rational trigonometric integrals. *Phys. D*, 280/281:59–72, 2014.

[285] R. Prohens and J. Torregrosa. New lower bounds for the Hilbert numbers using reversible centers. *Nonlinearity*, 32(1):331–355, 2019.

[286] M. Qi and L. Zhao. Bifurcations of limit cycles from a quintic Hamiltonian system with a figure double-fish. *Internat. J. Bifur. Chaos Appl. Sci. Engrg.*, 23(7):1350116, 15, 2013.

[287] J. Reyn. A bibliography of the qualitative theory of quadratic systems of differential equations in the plane, 1994.

[288] D. Rojas and J. Villadelprat. A criticality result for polycycles in a family of quadratic reversible centers. *J. Differential Equations*, 264(11):6585–6602, 2018.

[289] V. G. Romanovski and D. S. Shafer. *The center and cyclicity problems: a computational algebra approach.* Birkhäuser Boston, Ltd., Boston, MA, 2009.

[290] F. Rothe. The periods of the Volterra–Lotka system. *J. Reine Angew. Math.*, 355:129–138, 1985.

[291] R. Roussarie. On the number of limit cycles which appear by perturbation of separatrix loop of planar vector fields. *Bol. Soc. Brasil. Mat.*, 17(2):67–101, 1986.

[292] R. Roussarie. A note on finite cyclicity property and Hilbert's 16th problem. In *Dynamical systems, Valparaiso 1986*, volume 1331 of *Lecture Notes in Math.*, pages 161–168. Springer, Berlin, 1988.

[293] R. Roussarie. *Bifurcation of planar vector fields and Hilbert's sixteenth problem*, volume 164 of *Progress in Mathematics*. Birkhäuser Verlag, Basel, 1998.

[294] R. Roussarie and D. Schlomiuk. On the geometric structure of the class of planar quadratic differential systems. *Qual. Theory Dyn. Syst.*, 3(1):93–121, 2002.

[295] C. Rousseau. Hilbert's 16th problem for quadratic vector fields and cyclicity of graphics. *Nonlinear Anal.*, 30(1):437–445, 1997.

[296] C. Rousseau and H. Zhu. PP-graphics with a nilpotent elliptic singularity in quadratic systems and Hilbert's 16th problem. *J. Differential Equations*, 196(1):169–208, 2004.

[297] M. Sabatini. Characterizing isochronous centres by Lie brackets. *Differential Equations Dynam. Systems*, 5(1):91–99, 1997.

[298] J. A. Sanders and F. Verhulst. *Averaging methods in nonlinear dynamical systems*, volume 59 of *Applied Mathematical Sciences*. Springer-Verlag, New York, 1985.

[299] R. Schaaf. Global behaviour of solution branches for some Neumann problems depending on one or several parameters. *J. Reine Angew. Math.*, 346:1–31, 1984.

[300] D. Schlomiuk. Algebraic particular integrals, integrability and the problem of the center. *Trans. Amer. Math. Soc.*, 338(2):799–841, 1993.

[301] D. Schlomiuk, editor. *Bifurcations and periodic orbits of vector fields*, volume 408 of *NATO Advanced Science Institutes Series C: Mathematical and Physical Sciences*. Kluwer Academic Publishers Group, Dordrecht, 1993.

[302] D. Schlomiuk. Aspects of planar polynomial vector fields: global versus local, real versus complex, analytic versus algebraic and geometric. In *Normal forms, bifurcations and finiteness problems in differential equations*, volume 137 of *NATO Sci. Ser. II Math. Phys. Chem.*, pages 471–509. Kluwer Acad. Publ., Dordrecht, 2004.

[303] D. Schlomiuk, A. A. Bolibrukh, S. Yakovenko, V. Kaloshin, and A. Buium. *On finiteness in differential equations and Diophantine geometry*, volume 24 of *CRM Monograph Series*. American Mathematical Society, Providence, RI, 2005. Edited by Schlomiuk.

[304] D. Schlomiuk and N. Vulpe. Geometry of quadratic differential systems in the neighborhood of infinity. *J. Differential Equations*, 215(2):357–400, 2005.

[305] Y. Shao and Y. Zhao. The cyclicity and period function of a class of quadratic reversible Lotka–Volterra system of genus one. *J. Math. Anal. Appl.*, 377(2):817–827, 2011.

[306] Y. Shao and Y. Zhao. The cyclicity of a class of quadratic reversible system of genus one. *Chaos Solitons Fractals*, 44(10):827–835, 2011.

[307] J. Shi, W. Wang, and X. Zhang. Limit cycles of polynomial Liénard systems via the averaging method. *Nonlinear Anal. Real World Appl.*, 45:650–667, 2019.

[308] S. Shi. A concrete example of the existence of four limit cycles for plane quadratic systems. *Sci. Sinica*, 11:1051–1056, 1979. (In Chinese).

[309] S. Shi. A concrete example of the existence of four limit cycles for plane quadratic systems. *Sci. Sinica*, 23(2):153–158, 1980.

[310] K. S. Sibirskiĭ. On the number of limit cycles in the neighborhood of a singular point. *Differencial'nye Uravnenija*, 1:53–66, 1965.

[311] S. Smale. Dynamics retrospective: great problems, attempts that failed. *Phys. D*, 51(1-3):267–273, 1991. Nonlinear science: the next decade (Los Alamos, NM, 1990).

[312] S. Smale. Mathematical problems for the next century. *Math. Intelligencer*, 20(2):7–15, 1998.

[313] S. Sui and B. Li. Bounding the number of zeros of Abelian integral for a class of integrable non-Hamilton system. *Internat. J. Bifur. Chaos Appl. Sci. Engrg.*, 27(13):1750196, 9, 2017.

[314] S. Sui and L. Zhao. Bifurcation of limit cycles from the center of a family of cubic polynomial vector fields. *Internat. J. Bifur. Chaos Appl. Sci. Engrg.*, 28(5):1850063, 11, 2018.

[315] X. Sun and L. Zhao. Perturbations of a class of hyper-elliptic Hamiltonian systems of degree seven with nilpotent singular points. *Appl. Math. Comput.*, 289:194–203, 2016.

[316] G. Świrszcz. Cyclicity of infinite contour around certain reversible quadratic center. *J. Differential Equations*, 154(2):239–266, 1999.

[317] F. Takens. Forced oscillations and bifurcations. In *Applications of global analysis, I (Sympos., Utrecht State Univ., Utrecht, 1973)*, pages 1–59. Comm. Math. Inst. Rijksuniv. Utrecht, No. 3–1974. 1974.

[318] F. Takens. Forced oscillations and bifurcations. In H. W. Broer, B. Krauskopf, and G. Vegter, editors, *Global analysis of dynamical systems*, pages 1–61. Inst. Phys., Bristol, 2001.

[319] Y. Tang, D. Huang, S. Ruan, and W. Zhang. Coexistence of limit cycles and homoclinic loops in a SIRS model with a nonlinear incidence rate. *SIAM J. Appl. Math.*, 69(2):621–639, 2008.

[320] Y. Tang and W. Zhang. Bogdanov–Takens bifurcation of a polynomial differential system in biochemical reaction. *Comput. Math. Appl.*, 48(5-6):869–883, 2004.

[321] Y. Tang and W. Zhang. Heteroclinic bifurcation in a ratio-dependent predator-prey system. *J. Math. Biol.*, 50(6):699–712, 2005.

[322] Y. Tang and W. Zhang. Versal unfolding of planar Hamiltonian systems at fully degenerate equilibrium. *J. Differential Equations*, 261(1):236–272, 2016.

[323] Y. Tian and M. Han. Hopf and homoclinic bifurcations for near-Hamiltonian systems. *J. Differential Equations*, 262(4):3214–3234, 2017.

[324] Y. Tian and P. Yu. Bifurcation of ten small-amplitude limit cycles by perturbing a quadratic Hamiltonian system with cubic polynomials. *J. Differential Equations*, 260(2):971–990, 2016.

[325] Y. Tian and P. Yu. Bifurcation of small limit cycles in cubic integrable systems using higher-order analysis. *J. Differential Equations*, 264(9):5950–5976, 2018.

[326] C. Tung. Positions of limit cycles of the system $dx/dt = \sum_{0 \leq i+k \leq 2} a_{ik}x^i y^k$, $dy/dt = \sum_{0 \leq i+k \leq 2} b_{ik}x^i y^k$. *Sci. Sinica*, 8:151–171, 1959.

[327] A. N. Varchenko. Estimation of the number of zeros of an Abelian integral depending on a parameter, and limit cycles. *Funktsional. Anal. i Prilozhen.*, 18(2):14–25, 1984.

[328] M. Viano, J. Llibre, and H. Giacomini. Arbitrary order bifurcations for perturbed Hamiltonian planar systems via the reciprocal of an integrating factor. *Nonlinear Anal.*, 48(1, Ser. A: Theory Methods):117–136, 2002.

[329] J. Villadelprat. On the reversible quadratic centers with monotonic period function. *Proc. Amer. Math. Soc.*, 135(8):2555–2565, 2007.

[330] J. Villadelprat. The period function of the generalized Lotka–Volterra centers. *J. Math. Anal. Appl.*, 341(2):834–854, 2008.

[331] J. Villadelprat. On the period function in a class of generalized Lotka–Volterra systems. *Appl. Math. Comput.*, 216(7):1956–1964, 2010.

[332] J. Villadelprat. Bifurcation of local critical periods in the generalized Loud's system. *Appl. Math. Comput.*, 218(12):6803–6813, 2012.

[333] J. Villadelprat and X. Zhang. The Period Function of Hamiltonian Systems with Separable Variables. *J. Dynam. Differential Equations*, 32(2):741–767, 2020.

[334] M. Villarini. Regularity properties of the period function near a center of a planar vector field. *Nonlinear Anal.*, 19(8):787–803, 1992.

[335] J. Waldvogel. The period in the Lotka–Volterra system is monotonic. *J. Math. Anal. Appl.*, 114(1):178–184, 1986.

[336] J. Wang and D. Xiao. On the number of limit cycles in small perturbations of a class of hyper-elliptic Hamiltonian systems with one nilpotent saddle. *J. Differential Equations*, 250(4):2227–2243, 2011.

[337] J. Wang, D. Xiao, and M. Han. The number of zeros of Abelian integrals for a perturbation of hyperelliptic Hamiltonian system with degenerated polycycle. *Internat. J. Bifur. Chaos Appl. Sci. Engrg.*, 23(3):1350047, 18, 2013.

[338] N. Wang, J. Wang, and D. Xiao. The exact bounds on the number of zeros of complete hyperelliptic integrals of the first kind. *J. Differential Equations*, 254(2):323–341, 2013.

[339] N. Wang, D. Xiao, and J. Yu. The monotonicity of the ratio of hyperelliptic integrals. *Bull. Sci. Math.*, 138(7):805–845, 2014.

[340] C. Wu and Y. Xia. The number of limit cycles of cubic Hamiltonian system with perturbation. *Nonlinear Anal. Real World Appl.*, 7(5):943–949, 2006.

[341] J. Wu, L. Peng, and C. Li. On the number of limit cycles in perturbations of quadratic reversible center. *J. Aust. Math. Soc.*, 92(3):409–423, 2012.

[342] D. Xiao and Z. Zhang. On the uniqueness and nonexistence of limit cycles for predator-prey systems. *Nonlinearity*, 16(3):1185–1201, 2003.

[343] Y. Xiong and M. Han. New lower bounds for the Hilbert number of polynomial systems of Liénard type. *J. Differential Equations*, 257(7):2565–2590, 2014.

[344] S. Yakovenko. A geometric proof of the Bautin theorem. In *Concerning the Hilbert 16th problem*, volume 165 of *Amer. Math. Soc. Transl. Ser. 2*, pages 203–219. Amer. Math. Soc., Providence, RI, 1995.

[345] S. Yakovenko. Quantitative theory of ordinary differential equations and the tangential Hilbert 16th problem. In *On finiteness in differential equations and Diophantine geometry*, volume 24 of *CRM Monogr. Ser.*, pages 41–109. Amer. Math. Soc., Providence, RI, 2005.

[346] J. Yang and M. Han. Limit cycles near a double homoclinic loop. *Ann. Differential Equations*, 23(4):536–545, 2007.

[347] J. Yang, Y. Xiong, and M. Han. Limit cycle bifurcations near a 2-polycycle or double 2-polycycle of planar systems. *Nonlinear Anal.*, 95:756–773, 2014.

[348] J. Yang and P. Yu. Nine limit cycles around a singular point by perturbing a cubic Hamiltonian system with a nilpotent center. *Appl. Math. Comput.*, 298:141–152, 2017.

[349] J. Yang, P. Yu, and M. Han. Limit cycle bifurcations near a double homoclinic loop with a nilpotent saddle of order m. *J. Differential Equations*, 266(1):455–492, 2019.

[350] J. Yang and L. Zhao. Zeros of Abelian integrals for a quartic Hamiltonian with figure-of-eight loop through a nilpotent saddle. *Nonlinear Anal. Real World Appl.*, 27:350–365, 2016.

[351] J. Yang and L. Zhao. The cyclicity of period annuli for a class of cubic Hamiltonian systems with nilpotent singular points. *J. Differential Equations*, 263(9):5554–5581, 2017.

[352] J. Yang and L. Zhao. The perturbation of a class of hyper-elliptic Hamilton system with a double eight figure loop. *Qual. Theory Dyn. Syst.*, 16(2):317–360, 2017.

[353] X. Yang and Y. Ye. Uniqueness of limit cycle of the equation $dx/dt = -y + dx + lx^2 + xy + ny^2, dy/dt = x$. *J. Fuzhou Univ. Nat. Sci. Ed.*, 2:122–127, 1978. (In Chinese).

[354] Y. Ye. *Qualitative theory of polynomial differential systems*. Shanghai Scientific and Technical Publisher, Shanghai, 1995. (In Chinese).

[355] Y. Ye, S. Cai, L. Chen, K. Huang, D. Luo, Z. Ma, E. Wang, M. Wang, and X. Yang. *Theory of limit cycles*, volume 66 of *Translations of Mathematical Monographs*. American Mathematical Society, Providence, RI, second edition, 1986. Translated from the Chinese by Chi Y. Lo.

[356] J. Yu and C. Li. Bifurcation of a class of planar non-Hamiltonian integrable systems with one center and one-homoclinic loop. *J. Math. Anal. Appl.*, 269(1):227–243, 2002.

[357] P. Yu and M. Han. Twelve limit cycles in a cubic case of the 16th Hilbert problem. *Internat. J. Bifur. Chaos Appl. Sci. Engrg.*, 15(7):2191–2205, 2005.

[358] P. Yu and M. Han. A study on Zoladek's example. *J. Appl. Anal. Comput.*, 1(1):143–153, 2011.

[359] P. Yu and M. Han. Four limit cycles from perturbing quadratic integrable systems by quadratic polynomials. *Internat. J. Bifur. Chaos Appl. Sci. Engrg.*, 22(10):1250254, 28, 2012.

[360] P. Yu, M. Han, and J. Li. An improvement on the number of limit cycles bifurcating from a nondegenerate center of homogeneous polynomial systems. *Internat. J. Bifur. Chaos Appl. Sci. Engrg.*, 28(6):1850078, 31, 2018.

[361] P. Yu and F. Li. Bifurcation of limit cycles in a cubic-order planar system around a nilpotent critical point. *J. Math. Anal. Appl.*, 453(2):645–667, 2017.

[362] P. Yu and Y. Tian. Twelve limit cycles around a singular point in a planar cubic-degree polynomial system. *Commun. Nonlinear Sci. Numer. Simul.*, 19(8):2690–2705, 2014.

[363] X. Yu and X. Zhang. The hyperelliptic limit cycles of the Liénard systems. *J. Math. Anal. Appl.*, 376(2):535–539, 2011.

[364] H. Zang, M. Han, and D. Xiao. On Melnikov functions of a homoclinic loop through a nilpotent saddle for planar near-Hamiltonian systems. *J. Differential Equations*, 245(4):1086–1111, 2008.

[365] A. Zegeling and R. E. Kooij. The distribution of limit cycles in quadratic systems with four finite singularities. *J. Differential Equations*, 151(2):373–385, 1999.

[366] P. Zhang. On the distribution and number of limit cycles for quadratic system with two foci. *Acta Math. Sinica (Chin. Ser.)*, 44(1):37–44, 2001. (In Chinese).

[367] P. Zhang. On the distribution and number of limit cycles for quadratic systems with two foci. *Qual. Theory Dyn. Syst.*, 3(2):437–463, 2002.

[368] W. Zhang. Bifurcation of homoclinics in a nonlinear oscillation. *Acta Math. Sinica (N.S.)*, 5(2):170–184, 1989. A Chinese summary appears in Acta Math. Sinica **33** (1990), no. 3, 432.

[369] X. Zhang. The 16th Hilbert problem on algebraic limit cycles. *J. Differential Equations*, 251(7):1778–1789, 2011.

[370] X. Zhang. Inverse Jacobian multipliers and Hopf bifurcation on center manifolds. *J. Differential Equations*, 256(9):3278–3299, 2014.

[371] Z. Zhang. On the uniqueness of limit cycles of some nonlinear oscilation equations. *Dokl. Akad. Nauk SSSR*, 119:659–662, 1958. (In Russian).

[372] Z. Zhang. Proof of the uniqueness theorem of limit cycles of generalized Liénard equations. *Appl. Anal.*, 23(1-2):63–76, 1986.

[373] Z. Zhang, T. Ding, W. Huang, and Z. Dong. *Qualitative theory of differential equations*, volume 101 of *Translations of Mathematical Monographs*. American Mathematical Society, Providence, RI, 1992. Translated from the Chinese by Anthony Wing Kwok Leung.

[374] Z. Zhang and B. Li. High order Melnikov functions and the problem of uniformity in global bifurcation. *Ann. Mat. Pura Appl. (4)*, 161:181–212, 1992.

[375] Z. Zhang and C. Li. On the number of limit cycles of a class of quadratic Hamiltonian systems under quadratic perturbations. *Adv. in Math. (China)*, 26(5):445–460, 1997.

[376] Z. Zhang, C. Li, W. Li, and Z. Zheng. *An introduction to the bifurcation theory of vector fields*. Higher Education Press, Beijing, 1997. (In Chinese).

[377] L. Zhao. The perturbations of a class of hyper-elliptic Hamilton systems with a double homoclinic loop through a nilpotent saddle. *Nonlinear Anal.*, 95:374–387, 2014.

[378] L. Zhao. Zeros of Abelian integral for a class of quintic planar vector fields with a double homoclinic loop under non-symmetry perturbations. *Sci. Sin. Math.*, 47(1):227–240, 2017. (In Chinese).

[379] L. Zhao and D. Li. Bifurcations of limit cycles from a quintic Hamiltonian system with a heteroclinic cycle. *Acta Math. Sin. (Engl. Ser.)*, 30(3):411–422, 2014.

[380] L. Zhao, M. Qi, and C. Liu. The cyclicity of period annuli of a class of quintic Hamiltonian systems. *J. Math. Anal. Appl.*, 403(2):391–407, 2013.

[381] Y. Zhao. The monotonicity of period function for codimension four quadratic system Q_4^1. *J. Differential Equations*, 185(1):370–387, 2002.

[382] Y. Zhao. On the monotonicity of the period function of a quadratic system. *Discrete Contin. Dyn. Syst.*, 13(3):795–810, 2005.

[383] Y. Zhao. The period function for quadratic integrable systems with cubic orbits. *J. Math. Anal. Appl.*, 301(2):295–312, 2005.

[384] Y. Zhao, W. Li, C. Li, and Z. Zhang. Linear estimate of the number of zeros of Abelian integrals for quadratic centers having almost all their orbits formed by cubics. *Sci. China Ser. A*, 45(8):964–974, 2002.

[385] Y. Zhao, Z. Liang, and G. Lu. The cyclicity of the period annulus of the quadratic Hamiltonian systems with non-Morsean point. *J. Differential Equations*, 162(1):199–223, 2000.

[386] Y. Zhao and Z. Zhang. Abelian integrals for cubic vector fields. *Ann. Mat. Pura Appl. (4)*, 176:251–272, 1999.

[387] Y. Zhao and Z. Zhang. Bifurcations of limit cycles from cubic Hamiltonian systems with a center and a homoclinic saddle-loop. *Publ. Mat.*, 44(1):205–235, 2000.

[388] Y. Zhao and H. Zhu. Bifurcation of limit cycles from a non-Hamiltonian quadratic integrable system with homoclinic loop. In *Infinite dimensional dynamical systems*, volume 64 of *Fields Inst. Commun.*, pages 445–479. Springer, New York, 2013.

[389] Y. Zhao and S. Zhu. Perturbations of the non-generic quadratic Hamiltonian vector fields with hyperbolic segment. *Bull. Sci. Math.*, 125(2):109–138, 2001.

[390] Y. Zhou and C. Wang. The uniqueness of limit cycles for a quadratic system with an invariant straight line. *Int. J. Pure Appl. Math.*, 6(4):369–393, 2003.

[391] H. Zhu and C. Rousseau. Finite cyclicity of graphics with a nilpotent singularity of saddle or elliptic type. *J. Differential Equations*, 178(2):325–436, 2002.

[392] H. Żołądek. On a certain generalization of Bautin's theorem. *Nonlinearity*, 7(1):273–279, 1994.

[393] H. Żołądek. Quadratic systems with center and their perturbations. *J. Differential Equations*, 109(2):223–273, 1994.

[394] H. Żołądek. Eleven small limit cycles in a cubic vector field. *Nonlinearity*, 8(5):843–860, 1995.

[395] H. Żołądek. Remarks on: "The classification of reversible cubic systems with center" [Topol. Methods Nonlinear Anal. **4** (1994), no. 1, 79–136; MR1321810 (96m:34057)]. *Topol. Methods Nonlinear Anal.*, 8(2):335–342 (1997), 1996.

Printed in the United States
by Baker & Taylor Publisher Services